Soil Testing: Sampling, Correlation, Calibration, and Interpretation

Soil Testing: Sampling, Correlation, Calibration, and Interpretation

Proceedings of a symposium sponsored by Divisions S-4, S-8, A-1, A-4 and the Soil Testing and Plant Analysis Committee (S-877) of the American Society of Agronomy and Soil Science Society of America. The papers were presented during the annual meetings in Chicago, IL, 1–5 Dec. 1985.

Editor

J. R. Brown

Editorial Committee

J. R. Brown, *chair*   T. E. Bates   M. L. Vitosh

Symposium Planning Committee

R. D. Voss, *chair*   G. W. Randall
C. O. Plank   D. A.Whitney

Editor-in-Chief SSSA
John J. Mortvedt

Senior Managing Editor
Richard C. Dinauer

Assistant Editor
Kathryn A. Holtgraver

SSSA Special Publication Number 21

Soil Science Society of America, Inc.
Madison, Wisconsin, USA
1987

Cover Design: Julia M. Whitty

Soil Science Society of America, Inc.
677 South Segoe Road, Madison, WI 53711 USA

Library of Congress Cataloging-in-Publication Data

Soil testing.

    (SSSA special publication; no. 21)
    "The papers were presented during the annual meetings in
Chicago, IL, 1–5 Dec. 1985."
    1. Soils—Analysis. 2. Soil fertility. I. Brown, J. R. (James
Richard), 1931–   . II. American Society of Agronomy. III. Soil
Science Society of America. IV. Series.
S593.S7434 1987   631.4′22   87-3499
ISBN 0-89118-784-7

Printed in the United States of America

CONTENTS

Foreword vii

Preface ix

Conversion Factors for SI Units xi

1 Soil Sampling: Spatial and Temporal Variability
    Wayne E. Sabbe and David B. Marx 1

2 Soil Test Procedures: Correlation
    R.B. Corey 15

3 Soil Test Calibration
    Clyde E. Evans 23

4 Role of Response Surfaces in Soil Test Calibration
    Larry A. Nelson 31

5 Soil Testing Interpretations: Sufficiency vs. Build-up and
    Maintenance
    R.A. Olson, F.N. Anderson, K.D. Frank,
    P.H. Grabouski, G.W. Rehm, and C.A. Shapiro 41

6 Soil Test Interpretations: Basic Cation Saturation Ratios
    and Sufficiency Levels
    Donald J. Eckert 53

7 Soil Acidity: Soil pH and Lime Needs
    M.M. Alley and L.W. Zelazny 65

8 Status of Residual Nitrate-Nitrogen Soil Tests in the
    United States of America
    Gary W. Hergert 73

9 Sulfate: Sampling, Testing, and Calibration
    Gordon V. Johnson 89

10 Micronutrient Soil Tests: Correlation and Calibration
    F.R. Cox 97

11 Field Experimentation: Changing to Meet Current and
    Future Needs
    Malcolm E. Sumner 119

12 The Value and Use of Soil Test Summaries
    S.J. Donohue 133

FOREWORD

Developing the concepts of and formulating the scientific basis for soil testing has been one of the most important contributions made by soil scientists to the production of food and fiber. Much progress has been made toward the evaluation of the natural fertility of soils, understanding the mineral nutrition of plants, and increasing the efficiency of mineral fertilizers applied to the soil.

Many questions remain, however. Better analytical tools, refined experimental techniques, and new statistical tools continue to provide insights and interpretations that enhance the value of soil testing. New demands are placed on soil testing. Changes in farm management practices have been dramatic and have important consequences for soil testing. Reduced tillage requires a reevaluation of soil sampling procedures and soil test calibrations. The role of soil testing for evaluation of environmental pollution may be considered.

All these activities are being supported by traditional soil testing research. This book reports progress on many aspects of this research and attempts to satisfy the long standing need for a book setting forth the scientific principles upon which soil testing is based. The material in this book was first presented at a symposium sponsored by the Soil Testing and Plant Analysis Committee (S-877) and Divisions S-4, S-8, A-1, and A-4, held in conjunction with the annual meeting in Chicago, IL, 1-5 Dec. 1985.

May 1987

L. BOERSMA, *president,*
Soil Science Society of America

PREFACE

A special symposium on soil test calibration was held at the August 1975 American Society of Agronomy meeting in Knoxville, TN. The proceedings of that symposium were published in 1977 as ASA Special Publication no. 29 entitled *Soil Testing: Correlating and Interpreting the Analytical Results.*

The impact of the energy crisis of the 1970s, the grain embargo in response to USSR activities in Afghanistan, and other international and national activities all served to put a severe damper on export of U.S. agricultural products. This reduced the markets for agricultural products and the prices received. Environmental pressures have also mounted and there has been public concern for the protection of water quality and prevention of pollution from chemical fertilizers. The net result was a severe economic crisis in agriculture, which has placed increased emphasis on the efficient use of all agricultural inputs, including fertilizer. Soil testing plays a key role in the efficient use of fertilizer and its economic benefit.

Soil testing also plays an important role in the prevention of environmental degradation through providing guidelines to protect both surface, groundwater, and water quality.

A joint meeting between soil testing and plant analysis professionals from the Southeast and the North Central Regions in Memphis, TN, in October 1984 resulted in a resolution to update ASA Special Publication no. 29. The appeal was forwarded to the ASA planning committee.

This action resulted in a symposium held at the 1985 ASA meeting in Chicago, IL. This publication is the result of that symposium.

The Editorial Committee, the Planning Committee, and the many authors feel that much progress has been made in the 10 years since the Knoxville symposium. The interpretation of soil tests, however, often leads to confusion when the recommendations of different laboratories for samples from the same field are compared. This publication should reduce confusion surrounding interpretations of soil tests. Much remains to be done in sampling of soils and calibration of soil tests to ensure the efficient use of fertilizers in crop production.

January 1987

The Editorial Committee

J. R. BROWN, *chair, University of Missouri, Columbia Missouri*
T. E. BATES, *University of Guelph, Guelph, Ontario, Canada*
M. L. VITOSH, *Michigan State University, East Lansing, Michigan*

1 Soil Sampling: Spatial and Temporal Variability[1]

Wayne E. Sabbe and David B. Marx[2]

The knowledge of a soil's fertility status within a field permits the most reliable fertilizer management practices. The establishment and maintenance of a soil fertility program usually involves the use of a soil testing program. A complete soil testing program includes: (i) an accurate laboratory analytical procedure for estimating soil fertility (a service function), and (ii) appropriate correlation and calibration data for recommending the correct fertilization practices (a research function). Inherent in both the service and research functions is the assumption that the soil sample represents the field or plot. Vertical, horizontal, and time are three directions of variation that can occur when relying upon a soil sample to estimate soil fertility parameters. Variation in any of these sources can lead to an incorrect fertilizer or lime recommendation. Of equal or perhaps greater importance would be the variation from research data leading to recommendations. A recent trade magazine article (Jones, 1984) stated that perhaps over half of the soil tests are probably of little value because of improper soil sampling techniques. Therefore, the soil testing program is placed in jeopardy for both its analytical and recommendation roles due to nonrepresentative soil samples.

SOIL SAMPLING AND SOIL TESTING

A comparative study by the National Soil and Fertilizer Research Committee (Fitts et al., 1956) indicated that even the best selection of an extractant cannot predict 100% of the variation of plant growth vs. soil test values. This study, which involves 74 soils, 55 soil testing laboratories, and 10 extractants for soil phosphorus (P), produced a high correlation of 0.80 (i.e., a prediction of 64% of the variation in response). Therefore, a recommendation based on an excellent soil sampling procedure can be compromised by laboratory or interpretative shortcomings.

[1] Contribution from the Department of Agronomy and the Agricultural Statistics Laboratory, and published with the approval of the Director of the Agricultural Experiment Station, University of Arkansas-Fayetteville.

[2] Professor of Agronomy and Associate Professor of Statistics, respectively, University of Arkansas, Fayetteville, AR 72701.

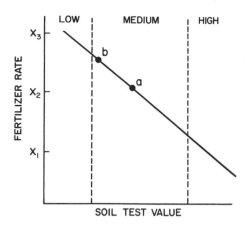

Fig. 1-1. Fertilizer response curve.

A redeeming factor that can negate variable or nonrepresentative soil samples is the noncontinuous function between soil test values and fertilizer rate (Fig. 1-1). This fertilizer recommendation curve is divided into low, medium, and high categories and any soil test value that falls into each category will receive a discrete fertilizer recommendation (X_1, X_2, or X_3). A soil sample taken from a field that was tested as *a* could contain some variation but still result in a correct fertilizer recommendation for the field. A field whose soil sample was tested as *b* would not necessarily receive a correct recommendation. Unless the soil sample can be assured of a true field representation, then differences in laboratory methods, fertilization philosophies, fertilizer technology, and computer adaptations can only be of academic interest.

The statistical approach to soil sampling has assumed that each observation is independent and identically distributed. Based on this assumption, several decades of statistical inferences concerning sampling have appeared in the literature. Cline (1944) presented general principles of soil sampling that were later expanded upon by Peterson and Calvin (1982). Generally the methods described involve statistical approaches such as the simple random, stratified random, systemic designs, and composite samples. These designs were based on the assumption that each observation was independent of other observations in the same field. Soil testing programs have relied on these assumptions and principles in formulating instructions for field sampling. Rigney (1956) illustrated the relative efficiency of five procedures for sampling where the fields were uniformity trials (Fig. 1-2). By using the value of method coefficient of variation (CV)/random CV and arbitrarily setting the random procedure's value at 1.00, Rigney concluded that the zigzag and stratified random methods were consistently superior to the random distribution. Because sampling by the zigzag method is a relatively easier method, its selection was the choice by many soil testing programs, including the Soil Test Work Group of the National Soil and Fertilizer Research Committee (Fitts et al., 1956).

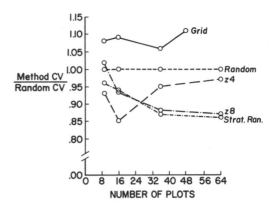

Fig. 1-2. Effect of plot number on randomization method (Rigney, 1956).

Reed and Rigney (1947) concluded in a soil testing program that, the field variation was much greater than laboratory variation, each soil property had an unique variation in a specific field, and the specific soil property having the greatest variation could not be anticipated. Therefore, the precision of field sampling is a major limiting factor in a soil testing program. An attempt to reduce the influence of variation was to introduce term(s) such as *allowable variation* (Keogh & Maples, 1967) or *maximum acceptable differences* (James & Dow, 1972). This term, one of arbitrary judgment, defined the acceptable difference between field mean and mean of samples such that a desired level of precision could be attained. Additionally, this term defined which soil test parameter had the greatest allowable variation and thereby determined the number of samples needed to describe the field. For example, Keogh and Maples (1967) determined that soil test P had the greatest allowable variation in nine of the 10 fields of alluvial soils. In other words, the number of samples needed to define the field mean P level was greater than the number of samples needed to define the other parameters. A corollary statement would be that the proper fertilizer recommendation could be achieved only when the number of samples necessary to give a correct P recommendation was achieved.

The size of the field did not affect the coefficient of variation appreciably, especially above a minimum size of 8.1 to 12.2 ha (20–30 acres) (Keogh & Maples, 1967). Cameron et al. (1971) reported that the number of samples needed to estimate the field average did not increase drastically with an increase in field size. Cypra et al. (1972) attempted to estimate the mean of a soil series by determining soil fertility parameters by horizon using seven fields 8 to 14 km apart. Some combinations of soil fertility and horizon would have required 175 fields; others only one to describe the series mean. If the A horizon was selected, then soil test P was the limiting fertility parameter. This combination required either 25 fields with 20 samples per field or 39 fields with one sample per field to determine the soil series mean for soil test P.

The deposition of dung and urine on grazed pastures increased the variation associated with soil sampling (Friesen & Blair, 1984). The most appropriate soil testing method appeared to be the cluster sampling (monitor plot), which allowed estimates of the temporal variations in soil test values.

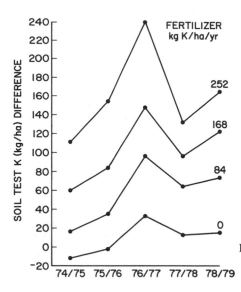

Fig. 1-3. Change in soil test K from fall to following spring in soil from fertilized plots (Petersen & Krueger, 1980).

Seasonal variability of soil fertility parameters has concerned soil testing programs because this variation can also result in an incorrect fertilizer and lime recommendation (Keogh & Maples, 1972; Liebhardt & Teel, 1977). Peterson and Krueger (1980) observed that soil test P and potassium (K) had cyclic variations with season and that the degree of these variations increased with an increase in the rate of P and K fertilizer (Fig. 1-3). A combination of soil test and more stable soil properties (i.e., texture, exchange acidity) did not reduce seasonal variation in the lime requirement of 11 Michigan field sites (Collins et al., 1970). Four of the 11 sites had a lime requirement when sampled in the spring, whereas all 11 had a lime requirement when sampled in mid-summer. From a standpoint of not recommending fertilizer when it is actually needed, Lochman and Molloy (1984) stated that if a preference existed, soil sampling in the fall would be preferable to obtaining spring soil samples. They also advocated the use of a moving average to reduce the coefficient of variation and standard deviation in estimating seasonal variations in soil test values.

SOIL SAMPLING INSTRUCTIONS

The previous discussion has illustrated several factors to consider in soil sampling for fertilizer and lime recommendations. Recently, Swenson et al. (1984) published guidelines for obtaining a soil sample that is to be used for measuring soil fertility parameters and fertilizer recommendations. The four areas for which instructions were given are: (i) accuracy and precision; (ii) sample areas that are representative of the field; (iii) effect of field size on accuracy; and (iv) when, how deep, and how often to sample. Data included in their report (Fig. 1-4 and 1-5) allowed for differences in precision, ac-

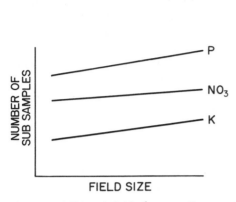
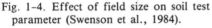

FIELD SIZE

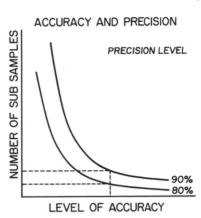

LEVEL OF ACCURACY

Fig. 1-4. Effect of field size on soil test parameter (Swenson et al., 1984).

Fig. 1-5. Relationship between precision and accuracy as affected by sampling number (Swenson et al., 1984).

curacy, field size, and nutrients. The timing, frequency, and depth of sampling depends on mobility of the nutrient. Mobile nutrients such as nitrate (NO_3) and sulfate (SO_4) should be estimated annually by soil sampling to depths such as 61 cm (24 in.), and samples need to be taken when biological activity is low. For immobile nutrients, a soil depth of 15 cm (6 in.) and a 3- to 5-year sampling interval was prescribed.

Conservation tillage demands a sampling technique to monitor the pH of the soil surface because of surface-applied nitrogen (N). Modified soil sampling instructions include obtaining a sample from the top 0- to 5- or 0- to 10-cm (0–2 or 0–4 in.) layer and from the traditional 0- to 15- or 0- to 20-cm (0–6 or 0–8 in.) layer. Mengel (1982) proposed the following sampling schemes for various conservation tillage methods: (i) where moldboard plowing is included, sample at least once every 4 years to plow depth; (ii) where tillage is reduced, or where N is injected below soil surface, sample at both the 0- to 10-cm and 10- to 20-cm (0–10 and 4–8 in.) depths; and (iii) where N is surface applied, sample at 0- to 5- and 5- to 20-cm depths (0–2 and 2–8 in.). A ridge-tilled field should probably be sampled after planting but before ridging (Moncrief et al., 1984). Samples taken at planting would not be compromised from soil thrown from the middles onto the ridges. These guidelines contain sampling aspects used by most soil testing programs to ensure representation of the field and reliability of the fertilizer and lime recommendation.

KRIGING APPROACH

In the agricultural sciences, the assumption of independent observations is not always met. In cases when the variables are spatially dependent, the usual classical statistical techniques are no longer valid. When sample data are collected from a field, the location of a sample to its neighbors may give some information as to the value obtained for that sample for the parameter being studied. Such data are spatially dependent. One method of handling spatially dependent variables is through the theory of regionalized variables developed by G. Matheron (1963) in the early 1960s. The application of this theory to problems originating in mining and geology led to the more popular name of geostatistics, of which "kriging" is a main branch. The problem of producing the "best" estimate of the unknown value of a parameter at some location within an ore deposit was termed *kriging* after D.G. Krige, who did much of the empirical work.

When a variable is distributed in space, it is said to be regionalized. A regionalized variable is a function, $f(x)$, that takes a value at every point, x_i, having coordinates (x_u, x_v, x_w) in three-dimensional space, where x_u represents the position of the point in the first dimension, x_v the position in the second dimension, and x_w the position in the third dimension.

The regionalized variable, which is usually quite irregular, possesses two characteristics: (i) a local random characteristic, and (ii) a general structured characteristic.

The variogram function is defined as the variance of the increment $(Z(x_i) - Z(x_j))]$ and is designated by

$$2P(x_i, x_j) = V[Z(x_i) - Z(x_j)].$$

For convenience we use half of the variogram, which is denoted by $P(h)$ and is called the *semivariogram*. Use of the semivariogram allows an easy interpretation. For example, when h is zero for all dimensions, $P(0)$ is the variability of the random variable $Z(x)$ when the point where the variable is sampled remains the same. A graph of a typical semivariogram is shown in Fig. 1-6. The distance at which samples become independent or are no longer correlated with each other is denoted by a and is called the "range of influence" or simply the "range." At this point, the semivariogram levels off at the "sill," denoted here by C or $C(0)$, since this is the covariance at distance 0. The sill should be approximately equal to the variance of the obser-

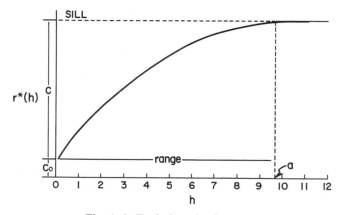

Fig. 1-6. Typical semivariogram.

vations taken from the field. Note that the units of the range are those of the random variable, whereas the units of the semivariogram are the squared units of the random variable.

The first step in estimating the semivariogram is to plot the data obtained from the equation

$$\varrho(h) = (1/2m) \sum_{i=1}^{m} \{[z(x_i) - z(x_i + h)]^2\} \qquad [1]$$

where m is the number of couples or pairs included in the summation. Because the spacing between locations of the observations will usually not differ by exactly h, all pairs of observations are compared and grouped by couples into intervals or lag classes according to the distance, h, they are apart. Thus, we have a data set consisting of the average distance that pairs are apart, the average value of the difference of the z's squared (and divided by 2), and the number of couples that went into this average.

For example, if soil samples are taken at regular intervals along a transect and analyzed for pH, then the semivariogram can be calculated from the resulting data. If these pH values are given in order as: 5.8, 5.7, 5.8, 6.0, 6.1, 5.9, 5.8, 5.8, 5.7, 5.5, 5.7; the semivariogram can then be calculated by the above formula in the following manner. For the first lag (pairs of observations that are one unit apart) the semivariogram is found by

$$(1/10) [(5.8-5.7)^2 + (5.7-5.8)^2 + (5.8-6.0)^2 + \cdots + (5.5-5.7)^2]$$

The second lag is calculated by

$$(1/9) [(5.8-5.8)^2 + (5.7-6.0)^2 + (5.8-6.1)^2 + \cdots + (5.7-5.7)^2].$$

In this example the lag classes with the respective semivariances are given in Table 1-1.

Table 1-1. Determination of semivariance for soil test pH.

	Data										
Distance, m	0	15.2	30.4	45.6	60.8	76.0	91.2	106.4	121.6	136.8	152
(ft)	(0)	(50)	(100)	(150)	(200)	(250)	(300)	(350)	(400)	(450)	(500)
pH	5.8	5.7	5.8	6.0	6.1	5.9	5.8	5.8	5.7	5.5	5.7

	Calculation	
n†	Log class	Semivariance
10	1	0.020
9	2	0.0433
8	3	0.0600
7	4	0.0614
6	5	0.850
5	6	0.860
4	7	0.450

† Numbers of pairs of observations used in the calculation.

Samples are usually collected on a regularly spaced rectangular or square grid. The spacing between sample locations, h, then takes on directional as well as distance properties. Usually four semivariograms can be calculated from a gridding of sample locations.

1. East-west (E-W)
2. North-south (N-S)
3. Northeast-southwest (NE-SW)
4. Northwest-southeast (NW-SE)

These directions are purely arbitrary and may not have any correspondence to actual N-S directions. It is possible that the semivariograms for the four directions will be different, indicating that different spatial variabilities exist depending upon the direction traveled (Fig. 1-7). A semivariogram based on data taken along a transect is valid only for changes in one direction along that transect. We have assumed throughout our discussion that the data are isotropic, i.e., that the semivariogram is the same in any direction. If the semivariograms calculated for the different directions are not the same, then anisotropy occurs and the calculations must be adjusted for its presence.

Once the semivariogram function has been estimated, the next step is to estimate the value of the dependent variable at the location, x. Assuming the semivariogram is now known, we may use a linear estimator, $Z(x)$, of the form

$$Z(x) = \sum_{j=1}^{n} f_j \, Z\,(x_j) \qquad [2]$$

where the $Z(x_j)$'s are the values of the observations and the f_j's are coefficients or weights associated with them. We wish for $Z(x)$ to be a minimum variance unbiased estimator.

Consequently, it is recommended that any samples lying outside the range of influence of the point to be kriged be excluded from the calculation. We have found that using the 20 closest samples to the point to be kriged works well. In addition, it is an appropriate technique to exclude samples until only positive weights are found.

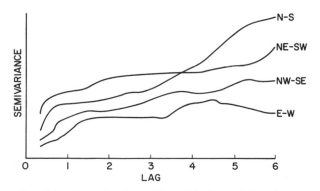

Fig. 1-7. Example of spatial variability and direction.

In soil sampling it is sometimes of major importance to find a precise and accurate estimate of a parameter for the entire field. Given a homogeneous field or a stratified part of a field, we will discuss two methods of sampling and present the results of a simulation study on the different sampling techniques. The sampling techniques studied are the simple random pattern and the zigzag pattern.

The simple random pattern is found by fixing the number of sample points to be taken and then randomly choosing the coordinates of the locations to be sampled. The zigzag pattern is obtained by choosing a starting portion of the field and a location randomly within this relatively small starting area. The other sample locations are then found by following a predetermined course from the original random starting point. This course is designed so that the entire field will be sampled. For example, if the area to be sampled is a rectangle, then the number of samples to be taken is determined and the area for a start defined. The zigzag pattern usually follows a course up or down the field with samples taken after a specified distance has been traversed in alternating NE-SW and NW-SE directions. The distance traveled between sample locations as well as the actual direction depends upon the size and shape of the field to be sampled.

Figures 1–8 and 1–9 show a zigzag pattern for a rectangular field and a triangular field. The starting area is indicated in both fields, where 10

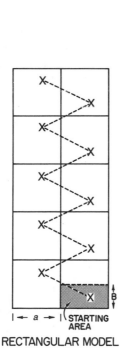

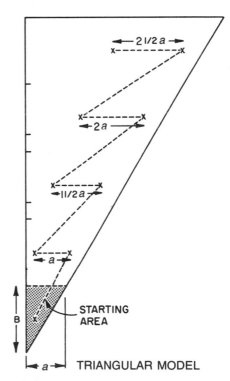

Fig. 1–8. Rectangular model for zigzag sampling.

Fig. 1–9. Triangular model for zigzag sampling.

samples apiece were to be taken. The rectangular field pattern was determined by deciding that only one zigzag up the field was necessary to adequately sample it. Thus there should be five points on each side of the field. Blocking the field gives the realization that there will be a sample in each of the 10 defined blocks. In addition, since each block has dimensions a and $2B$, the distance and direction for the zigzag pattern is defined. The distance is $(a^2 + B^2)^{1/2}$ and the direction is given by $\theta = \text{tangent } (B/a)$. Once the sampling plan has been devised, the actual samples may be collected by the shortest route. For example, in Fig. 1–8, one would simply collect all the samples on the east side of the field and then collect the western half of the samples on the return trip.

For the triangular field, the easiest way to form a zigzag pattern that covers the field is to again determine the number of samples desired. Take one-half this number and divide it into the length of the field. In our example we have 10 samples; hence, the length is divided into five parts. The starting area is the bottom portion, which is one-fifth of the entire length. The zigzag pattern then evolves by defining the first direction as $\theta = \text{tangent } (B/a)$ and distance as $(a^2 + B^2)^{1/2}$. The third location is distance a from the second. The fourth location is $2a$ to the east and B north from the third location. Location five is west $1\frac{1}{2}a$. Continue the same sequence traveling north B and east the same distance as on the preceding westward move, but move an additional a east. The difference between the rectangular and the triangular fields lie in the direction and distance to the next sample location. The triangular field uses varying distances and directions as seen in Fig. 1–9. If the field is too wide, the zigzag is repeated down the other side of the field. Since this pattern tends to undersample the wider portion of the field, an alternative approach is to use the same triangular scheme as described above to determine the right-hand side sample location but to then have one, two, two, three, four, four, five, six, six, etc., samples in each row all $1.5a$ units apart. This scheme is depicted in Fig. 1–10.

A simulation study was run with three fields using soil test pH data and two fields using soil test K data. Both K fields were rectangular, as was one pH field, the other two pH fields were triangular. One thousand samples of size five were simulated using the procedures just described. The sample data were found by kriging these sample locations. The results are given in Tables 1–2 and 1–3.

Although both methods are theoretically unbiased, the zigzag pattern may be slightly biased once the starting areas have been defined. This bias, if it occurs, is offset by the consistency of the means of the groups of five data points and by ensuring that the entire field is sampled. This consistency is demonstrated where the maximum and minimum values for the mean of five points over the one thousand simulations are shown. The completely random sampling scheme generally has a much larger range of mean values than does the zigzag pattern. Consequentially, data collected in a zigzag pattern will lead to more consistent soil test recommendations than data collected in a completely random fashion.

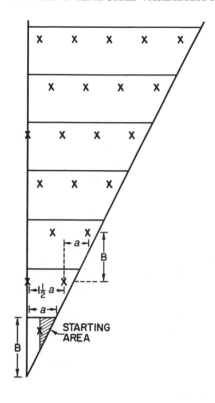

Fig. 1-10. Alternate triangular model for zigzag sampling.

Table 1-2. Simulation soil test pH.†

Site	Sampling	Mean	Max	Min	Variance
Triangular	Random	5.98	6.38	5.73	0.013
(5 points)	Zigzag	5.80	5.97	5.65	0.006
Rectangular	Random	5.69	5.78	5.59	0.004
(5 points)	Zigzag	5.77	5.81	5.75	0.026

† $n = 1000$.

Table 1-3. Simulation soil test K.†

Site	Sampling	Statistical parameters			
		Mean	Max	Min	Variance
			mg kg^{-1}		
Rectangular	Random	1310	1990	945	610
(5 points)	Zigzag	1560	1890	1050	1610
Rectangular	Random	1335	1680	1095	280
(10 points)	Zigzag	1320	1580	945	475

† $n = 1000$.

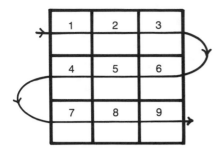

Fig. 1-11. Strip sampling technique (Assmus et al., 1985).

When the stimulation is modified so that the sample size is 10 instead of five, the difference in the means of the two sampling methods is decreased. The range is also reduced. Again, the importance of the number of samples must be emphasized.

We again emphasize that both methods of sampling are essentially unbiased. Indeed, even if the chosen sample locations are close enough together that they are spatially dependent, the procedures will yield unbiased estimates of the mean value for the field. However, the variance of the mean based upon spatially dependent observations is biased unless this spatial dependency is taken into account. Generally, we are not too concerned about this bias in the variance if the field is uniformly sampled and the sole reason for sampling is for a soil test recommendation. However, if inferences about the field are to be made based upon the sample, then adjusting for the bias in variance is extremely important.

An adaption of kriging was presented by Assmus et al. (1985). Semivariograms for soil test P at three sites revealed spatial correlations for two sites and a pure nugget effect for the third site. A strip sampling technique (Fig. 1-11) was used to detect the spatial variability within these fields, which showed a decrease of the number of samples required. The strip sampling (horizontal and vertical) revealed the pattern of variability, thereby modifying the fertilizer recommendation as to the direction of application.

The problem of anisotropy can be handled mathematically and can actually be used to help design plot shape and direction, as suggested by McBratney (1985). The smaller plot dimension should be situated in the direction of maximum variation with the dimensions of the plot being identical to the anisotropy ratio.

For example, assume a field area that was to contain the research area had a N-S semivariance value of 5, whereas the E-W semivariance value is 2. Therefore, the smaller plot dimension should be in a N-S direction with the greater plot dimension in the E-W direction. Ideally the ratio of the two dimensions (N-S/E-W) should be 2:5.

CURRENT STATISTICAL CONCERNS

Recent technological changes in farming practices and fertilizer applications have complicated obtaining a representative soil sample. Row crop agriculture has initiated a "stale bed" format where successive crops are grown on preexisting beds. This concept allows for fertilizer application to be placed consecutively in approximately the same location. Therefore, a bed profile would show fertilized and nonfertilized zones confounding the task of obtaining a soil sample that represents its actual fertility status. Fertilizer applicators have the ability to band-apply nutrients for more effective and economic yield response. Again, these variable soil fertility zones cause problems with soil sampling procedures.

Conservation tillage fertilizer practices have resulted in nutrient accumulation in the upper soil region. Broadcast N applications have decreased soil pH in these regions causing a vertical variation in soil fertility. An incorrect sampling depth may disguise the fact that the upper root zone [i.e., 5–6 cm (2–3 in.)] soil sample would have fertility status different from a routine 15- to 20-cm (6–8 in.) soil sample.

Recent studies on fertilizer efficiency indicate that yields are affected by the fraction of soil fertilized. The resultant model (Barber, 1984) takes into consideration the soil solution, the buffer power of soil, and the rate of diffusion. From these considerations, the model is useful in determining the most efficient fertilizer placement (i.e., the volume of soil to fertilizer mix). As the concept of fertilizer efficiency and fertilizer application technology advance, fertilizer placement will cause variation (vertical and horizontal) in soil fertility, thereby comfounding the soil sampling procedure.

The above-mentioned considerations will challenge our ability to represent a field with a single soil sampling procedure. Each field will present a set of parameters that will dictate a specific sampling pattern. The challenge will be to define the set of parameters so that the number of samples taken will estimate the mean in spite of the spatial and temporal variability.

SUMMARY

Accurate limestone and fertilizer recommendations require a soil sample that is representative of the field. This representation should be acquired with the least number of subsamples to lower the time and cost per recommendation. Increasing the number of subsamples increases the precision and accuracy but not in a linear fashion; therefore, the level of precision and accuracy needed will dictate the number of subsamples. Studies have indicated that field size, tillage method, and nutrient (i.e., depth and frequency) affect the number of subsamples per field. The use of kriging has the ability to define the direction and magnitude of variability within a field using a minimum of subsamples. Kriging may have its greatest advantage in the determination of plot size and shape in selection of field research areas. While random or zigzag sampling is advocated for farmers' fields, other sampling patterns can aid in defining field variation and therefore assist in the resultant fertilization practices.

REFERENCES

Assmus, R.A., P.E. Fixen, and P.D. Evenson. 1985. Detection of soil phosphorus spatial variability through the use of semivariograms and strip sampling. J. Fert. Issues 2:136-143.

Barber, S.A. 1984. Soil nutrient bioavailability: A mechanistic approach. John Wiley & Sons, Inc., New York.

Cameron, D.R., M. Nybert, and J.A. Toogood, and D.H. Loverty. 1971. Accuracy of field sampling for soil tests. Can. J. Soil Sci. 51:165-175.

Cline, M.G. 1944. Principles of soil sampling. Soil Sci. 58:275-288.

Collins, J.B., E.P. Whiteside, and C.E. Cress. 1970. Seasonal variability of pH and lime requirements in several southern Michigan soils when measured in different ways. Soil Sci. Soc. Am. Proc. 34:56-61.

Cypra, J.E., O.W. Bidwell, D.A. Whitney, and A.M. Feyerherm. 1972. Variations with distance in selected fertility measurements of pedons of western Kansas Ustoll. Soil Sci. Soc. Am. Poc. 36:111-115.

Fitts, J.W. et al. 1956. Soil tests compared with field, greenhouse, and laboratory results: A comparative study. North Carolina Agric. Exp. Stn. Tech. Bull. 121.

Friesen, D.K., and G.J. Blair. 1984. A comparison of soil sampling procedures used to monitor soil fertility in permanent pastures. Aust. J. Soil Res. 22:81-90.

James, D.W., and A.I. Dow. 1972. Source and degree of soil variation in the field: The problem of sampling for soil tests and estimating soil fertility status. Washington State Univ. Agric. Exp. Stn. Bull. 749.

Jones, J.B., Jr. 1984. Soil tests: Are they really good? Agrichemical Age 28(4):33-34.

Keogh, J.L., and R. Maples. 1967. A statistical study of soil sampling of Arkansas alluvial soils. Arkansas Agric. Exp. Stn. Rep. Series 157.

Keogh, J.L., and R. Maples. 1972. Variations in soil test results as affected by seasonal sampling. Arkansas Agric. Exp. Stn. Bull. 777.

Liebhardt, W.C., and M.R. Teel. 1977. Fluctuations in soil test values for potassium as influenced by time of sampling. Commun. Soil Sci. Plant Anal. 8:591-597.

Lockman, R.B., and M.G. Molloy. 1984. Seasonal variation in soil test results. Commun. Soil Sci. Plant Anal. 15:741-757.

Matheron, G. 1963. Principles of geostatistics. Econ. Geol. 58:1246-1266.

Mengel, D.B. 1982. Developing fertilizer programs for conservation tillage. In Proc. Indiana Plant Food and Agric. Chem. Conf., West Lafayette, IN. 14-15 December. Purdue University, West Lafayette, IN.

McBratney, A.B. 1985. The role of geostatistics in the design and analysis of field experiments with reference to the effect of soil properties on crop yield. p. 3-8. In D.R. Nielson and J. Bouma (ed.) Soil spatial variability. Proc. Workshop of the ISSS and SSSA, Las Vegas. 30 Nov.-1 Dec. 1984. Centre for Agricultural Publishing and Documentation (PUDOC), Wageningen, Netherlands.

Moncrief, J.F., W.E. Fenster, and G.W. Rehm. 1984. Effect of tillage on fertilizer management. p. 45-56. In Conservation tillage for Minnesota. University of Minnesota Agric. Ext. Serv. Publ. AG-BU-2402.

Peterson, L.A., and A.R. Krueger. 1980. Variation in content of available P and K (Bray I) in soil samples from a cropped N, P, and K fertility experiment over 8 years. Commun. Soil Sci. Plant Anal. 11:993-1004.

Peterson, R.G., and L.D. Calvin. 1982. Sampling. In A. Klute (ed.) Methods of soil analysis, Part 1. 2nd ed. Agronomy 9:33-51.

Reed, J.F., and J.A. Rigney. 1947. Soil sampling from fields of uniform and nonuniform appearance and soil types. Agron. J. 39:26-40.

Rigney, J.A. 1956. Sampling soils for composition studies. Proc. Am. Soc. Hort. Sci. 68:569-575.

Swenson, L.J., W.C. Dahnke, and D.D. Patterson. 1984. Sampling for soil testing. North Dakota State University, Dep. of Soil Sci., Res. Rep. no. 8.

2   Soil Test Procedures: Correlation[1]

R. B. Corey[2]

The development of a soil test for a given nutrient has historically involved three steps: (i) selecting an extractant, (ii) correlating the amount of nutrient extracted (test value) with the amount taken up by the plant, and (iii) calibrating the test value in terms of its effect on some desirable crop characteristic, usually yield of marketable product. Fertilizer recommendations are then based on interpretation of calibration data and fertilizer response curves. This chapter is concerned primarily with the correlation step, but the proposed quasimechanistic approach to the correlation process necessarily involves a discussion of the soil variables that should be measured to provide input to mathematical models describing nutrient uptake by roots.

NUTRIENT UPTAKE—THE STANDARD FOR COMPARISON

Nutrients extracted in soil testing procedures are frequently referred to as available nutrients. What is meant by the term, available, is not always clear. The *Glossary of Soil Science Terms* (SSSA, 1987) defines *available nutrients* as "Nutrient ions or compounds in forms which plants can absorb and utilize in growth." No mention is made of the rate at which the available nutrient can be absorbed. For purposes of soil test correlation, the rate at which a nutrient is taken up is more important than the size of the available nutrient pool. Ideally, soil test results, combined with other pertinent information, should enable one to predict the total uptake of a given nutrient by a given crop over a growing season or during a critical growth period if crop yield is not limited severely by other factors. Thus, correlation of estimated nutrient uptake, based on the soil test and any other pertinent factors, with total uptake by the crop would seem to be a logical approach. If nutrient concentration in a specific plant part at a specific growth stage is highly correlated with total uptake, the usual case when yield differences are minimal, this more easily obtained property could be substituted for total uptake.

---

[1] Contribution from the College of Agricultural and Life Sciences, University of Wisconsin-Madison.

[2] Professor, Department of Soil Science, University of Wisconsin-Madison, Madison, WI 53706.

   Correlation studies can be performed in the field. However, growth chamber conditions offer better control of environmental factors and soil homogeneity. Also, the small volumes of soil required make studies utilizing large numbers of samples feasible. Application of greenhouse results to field conditions, however, may be suspect because of the different conditions, particularly temperature, radiation intensity, root density, and transpiration rate, which affect the relative importance of diffusion and mass flow. Careful selection of environmental conditions in the growth chamber and harvesting plants before the roots form mats at the edges of the containers can minimize, but may not eliminate, these problems.

   In field studies, soil samples should be taken from the root zone of the plants that are sampled for uptake. Sampling soil prior to planting is preferable. If tissue samples (e.g., ear leaf) are taken as an estimate of total uptake, soil samples taken from the root zone of the sampled plants at the time of sampling should correlate well with preplant samples if the amount of nutrient extracted at that growth stage is small compared with the labile pool.[3] This is the usual case for soil phosphorus (P), but not for potassium (K) on soils with low buffer power.[4]

THE CORRELATION PROCESS

   Nutrient concentrations in soil extracts can be correlated directly with plant uptake, but low correlation coefficients usually result if soils with widely differing characteristics are included in the study. Correlations can almost always be improved by grouping soils with similar characteristics and correlating (and calibrating) the test within the groups. An alternative (or addition) to this approach utilizes multiple regression analysis as a means of including effects on uptake by other measured variables such as pH, texture, and organic matter.

   In using multiple regression analysis, the researcher produces a model of the uptake process that tells how the included variables contribute to uptake of the nutrient being studied. For example, in the regression equation,

$$U = a + b_1 X_1 + b_2 X_2,$$ [1]

if U = P uptake, X_1 = soil test P, and X_2 = pH, a unit change in soil test P changes U by a factor of b_1, and a unit change in pH changes U by a factor of b_2. Inclusion of pH in this manner may increase the correlation between determined and predicted uptake but mechanistically the model makes no sense. According to the equations, P uptake will vary with pH even if

---

   [3] The term, *labile pool*, is defined (SSSA, 1987) as "The sum of an element in the soil solution and the amount of that element readily solubilized or exchanged when the soil is equilibrated with a salt solution." As used in this chapter, the labile pool includes all forms that are exchangeable with added isotopes of a particular element.

   [4] The term, *buffer power*, is defined as the change in the concentration of labile nutrient per unit change in concentration of dissolved nutrient (Nye & Tinker, 1977, p. 42).

no P is there! Inclusion of additional variables may further improve the correlation, but the relationship derived has no mechanistic basis and therefore applies only to the sample population used.

This problem can be partially solved by substituting the logarithms of the variables for the variables in Eq. [1] and deriving different regression coefficients, a', b_1', and b_2'

$$\log U = \log a' + b_1' \log X_1 + b_2' \log X_2. \qquad [2]$$

This is equivalent to

$$U = a' X_1^{b_1'} X_2^{b_2'}. \qquad [3]$$

This produces an interactive model in which uptake approaches zero as soil test P approaches zero regardless of pH. Intuitively, this equation makes more sense, and there is some theoretical justification based on $OH^- - H_2PO_4^-$ competition for adsorption sites, but the relationship derived from the regression equation is still largely empirical. It remains empirical no matter how many independent variables are added in an attempt to improve the correlation.

An empirical, stochastic approach to soil test correlation is not necessarily bad. On the contrary, it has been the logical approach to use in the absence of detailed mechanistic models for plant-nutrient uptake. Now, however, such models are available (Nye & Tinker, 1977; Barber, 1984), and we should use them to guide the development and correlation of improved testing methods.

QUASIMECHANISTIC MODELS

A comprehensive solute transport model describes the flux of ions in the soil solution to root absorbing surfaces by diffusion in response to a concentration gradient and by convection caused primarily by plant transpiration. At usual concentrations in the soil solution, diffusion has been found to be the dominant transport mechanism for nutrients other than calcium (Ca^{2+}), magnesium (Mg^{2+}), and sulfate (SO_4^{2-}), thus the subsequent discussion will center on this process.

Soil factors that affect diffusive transport include water content, nutrient concentration in solution, and the ability to resupply absorbed nutrients (buffer power). Important plant factors include root geometry (root radius, presence of root hairs/mycorrhizae) and root uptake physiology (root absorbing power). How these factors interact is shown in Eq. [4], which is a modification of an uptake equation derived by Baldwin et al. (1973) that describes the diffusive radial flux of nutrients from an isotropic medium (soil) to a cylindrical sink (plant root), assuming depletion of a cylindrical volume of soil surrounding each segment of root.

$$U = C_{li}\, b \left[1 - \exp \frac{-2\pi\, \alpha A_l r_o L_v t}{b \left(1 + \dfrac{\alpha A_l r_o}{D_l\, \theta f} \ln \dfrac{r_h}{1.65\, r_o} \right)} \right] \qquad [4]$$

where soil factors
  C_{li} = initial concentration of nutrient in soil solution,
  b = buffer power,
  A_l = fractional area of water contacting root,
  θ = volumetric water content,
  D_l = diffusion coefficient in soil solution, and
  f = conductivity factor related to diffusion pathlength; and
where plant factors
  U = uptake per unit volume of soil in time, t,
  α = root absorbing power,
  r_o = root radius,
  r_h = half-distance between roots, and
  L_v = root density.

The conductivity factor, f, decreases with a decrease in θ because of greater tortuosity of the diffusion path at lower water contents. The fractional area of water contacting the root, A_l, is numerically equal to θ. The buffer power, b, is equal to the change in concentration of total labile nutrient per unit change in concentration of that nutrient dissolved in the soil solution. The labile form includes dissolved and reactive adsorbed forms. Soils with high adsorption capacities for specific nutrients generally show high buffer powers for those nutrients.

The root density, L_v, is equal to the length of root per unit volume of soil. The value of L_v is readily determined for roots without root hairs or mycorrhizae, but their presence makes the geometry of the nutrient absorbing system much more difficult to describe quantitatively. The root absorbing power, α, is equal to the uptake flux density divided by the nutrient concentration at the root surface. In some cases, α can be described by a Michaelis-Menten-type plot of flux density vs. concentration at the root surface. However, this relationship has been shown to depend on the pre-existing nutrient status of the plant.

At first glance, even this simplified equation appears to be extremely complex for application to routine soil testing. However, it does point out the soil and plant variables that must be measured or estimated. Of the soil variables, C_{li}, b, and θ are critical. The plant variables, α, r_o, and L_v are responsible for differences in uptake from the same soil by different species or cultivars.

In a routine soil testing program, it is not feasible to measure all of the relevant soil variables. For many nutrients, however, measurements of C_{li} and estimations of b, θ, and f are feasible. The plant variables that enter into the equation are difficult to measure, but these measurements may not be necessary if the plant characteristics are similar in different soils. For example, Eq. [4] can be rewritten as

$$U = C_{li}\, b \left\{ 1 - \exp \left[\frac{-2\pi\, A_l P_n}{b\left(1 + \dfrac{A_l}{D_l \theta f}\, P_d\right)} \right] \right\} \qquad [5]$$

where P_n = product of plant and time factors in the numerator of Eq. [4], and P_d = product of plant factors in the denominator. If $1 \gg A_l P_d / D_l \theta f$, then

$$U = C_{li} b\, [1 - \exp(-2\pi\, A_l P_n / b)]. \qquad [6]$$

If $1 \ll A_l P_d / D_l \theta f$, then

$$U = C_{li} b\, [1 - \exp(-2\pi\, D_l \theta f\, P_n / b P_d)]. \qquad [7]$$

In either of these cases, P_n or P_n/P_d can be estimated stochastically by using measured or estimated values of D_l, θ, f, and b and varying P_n or P_n/P_d over a range that makes the value of the exponent vary from about -0.01 to -5, then determining the value of P_n or P_n/P_d that results in the highest correlation coefficient when estimated uptakes are correlated with measured values. The range in exponent values suggested causes $[1 - \exp(-2\pi\, D_l \theta f\, P_n/b P_d)]$ in Eq. [7] to range from 0.01 to 0.99. In some cases, it might be necessary to extend the upper range of the exponent to more nearly approach zero.

If $1 \cong A_l P_d / D_l \theta f$, the P terms cannot be combined into a single variable. Therefore, in seeking the best values through a correlation approach, P_d must be varied systematically so that $A_l P_d / D_l \theta f$ ranges from a minimum of 0.1 to a maximum of 10, assuming 0.1 and 10 are selected to satisfy the criteria that 1 is much greater or less than $A_l P_d / D_l \theta f$, respectively. At each value of P_d, P_n is varied such that the value of the exponent term ranges from -0.01 to -5. The values of P_n and P_d selected are those that give the highest correlation between calculated and measured uptake.

In deciding on which approach to take, the latter one should be tried first. If a single value of P_n correlates best regardless of the value assigned to P_d, Eq. [6] applies. If the highest correlation is achieved at a specific value of P_n/P_d regardless of absolute values, Eq. [7] applies. Otherwise the bivariate approach must be used. The success of this approach depends on how applicable Eq. [4] is for predicting nutrient uptake by a particular crop, the constancy of the P values from site-to-site, and the degree to which the analytical methods give accurate estimates of C_{li}, b, and θ. The ultimate test is the correlation achieved between predicted uptake and uptake by field-grown crops.

A preliminary attempt was made at testing this approach in a study concerned with availability of soil zinc (Zn) to 2- and 3-year-old red pine (*Pinus strobus*) seedlings (J. Zhang, J.G. Iyer, & R.B. Corey, unpublished data). A total of 26 samples of both soil and needles were taken from three tree nurseries. Measurements of C_{li} and b were obtained by a modification of the method of Fujii et al. (1983). To determine solution Zn, 10 g of soil was equilibrated for 30 min with a solution containing 0.01 M Ca(NO$_3$)$_2$ and 5 mg Cd^{2+} L^{-1}. Total labile Zn was determined by extraction with 0.01 M Ca(NO$_3$)$_2$ + 0.005 M EDTA (ethylenediaminotetraacetic acid). The Cd was included in the solution so that adsorbed Zn could be estimated from the following equation, thereby eliminating the need for a second extraction,

$$[Zn_{ads}] = k_{Cd}^{Zn} [Cd_{ads}](Zn^{2+})/(Cd^{2+}).$$ [8]

The brackets in Eq. [8] denote concentrations and parentheses denote activities. The selectivity coefficient, k_{Cd}^{Zn}, used was the average value determined for all soils assuming that the ratio $(Zn^{2+})/(Cd^{2+})$ was equal to the ratio of the total concentrations in solution and that $[Cd_{ads}]$ was equal to total Cd added, minus that remaining in solution.

In this study, all of the soils were sandy and had similar values of θ. Therefore, θ, A_l, and f were considered to be constant and the following equation was used:

$$U = C_{li}b[1 - \exp(-P/b)]$$ [9]

where

$$P = \frac{2\pi \, \alpha A_l r_o L_v t}{1 + \dfrac{\alpha A_l r_o}{D_l \theta f} \ln \dfrac{r_h}{1.65 \, r_o}}.$$ [10]

Concentration of Zn in the needles was assumed to be proportional to U and was used in the correlations. The value of P was varied from 0 to 80 and the correlation coefficient between measured Zn concentrations in needles and calculated Zn uptake was plotted as a function of P. The plot showed a sharp maximum at $r = 0.80$ and $P = 15$. The correlation coefficients for the relationships of Zn in needles with total labile Zn and with solution Zn were 0.69 and 0.66, respectively. Thus, this quasimechanistic approach increased the prediction of Zn concentration in needles considerably ($R^2 = 0.64$ compared with 0.48 and 0.44 using labile and solution Zn, respectively).

Wietholter (1983) used a somewhat different quasimechanistic approach in estimating K availability to corn (*Zea mays* L.). He assumed that each centimeter of root at one time had root hairs active for 5 days, and that soil within the cylinder containing root hairs was rapidly depleted of K after which K diffused from the surrounding soil to the root-hair cylinder. He further assumed that older sections of roots without active root hairs did not participate in K uptake because of the length of time required to backfill the depleted zone formed when the root hairs were active. He used Eq. [4] along with literature values for root-hair length, L_v, and α to calculate the time required for depletion of K from the root-hair cylinder, and a diffusion equation assuming constant root absorbing power (Nye & Tinker, 1977) to calculate K flux to the root-hair cylinder after depletion within the cylinder. A solution of 0.004 M $Sr(NO_3)_2$ was used as the equilibrating solution at a solution/soil ratio of 2.5. Concentrations of strontium (Sr), Ca, K, sodium (Na), and manganese (Mn) in solution were determined with the inductively coupled plasma emission spectrometer (ICP), and calculations of exchangeable values and buffer powers were made based on the partitioning of the added Sr between adsorbed and dissolved phases and predetermined selectivity coefficients (Gapon K's) for the competitive adsorption reactions among the various cations.

Correlation coefficients for the relationships between K concentration in corn ear leaves and calculated K uptake, solution K, and exchangeable K, respectively, were 0.76, 0.70, and 0.38 using 51 samples of soil and leaf tissue from a variety of Wisconsin soils ranging in texture from sands to silty clay loams. This converts to R^2 values of 0.58, 0.49, and 0.14, respectively.

SUMMARY

Mathematical models describing the kinetics of nutrient uptake by plants from soils indicate the soil and plant variables that affect the uptake process. Although measuring each of these variables is not feasible in a routine soil testing program, it should be possible to design cost-effective methods for determining key variables and estimating the others.

The models should be particularly helpful in designing quasimechanistic approaches to soil test correlation. The approaches outlined in this paper are illustrative of the way in which mechanistic and stochastic methods can be integrated in designing correlation procedures. Results of the preliminary studies presented here are encouraging, but much development work remains to be done. Uptake equations other than Eq. [4] may prove to be more effective in predicting nutrient uptake. The possibility that P_n or P_n/P_d, variables incorporating the products of the plant factors in Eq. [5], have characteristic values for specific crops merits research. Also, desorption studies might show that soil series or soil-texture groups are homogeneous enough to use one curve for relating solution concentration to total labile nutrient to define b as a function of C_l for the series or group. The research needs are great, but the models provide a focus for the research and a framework to integrate results from different researchers.

REFERENCES

Baldwin, J.P., P.H. Nye, and P.B. Tinker. 1973. Uptake of solutes by multiple root systems from soil. III. A model for calculating the solute uptake by a randomly dispersed root system developing in a finite volume of soil. Plant Soil 38:621–635.

Barber, S.A. 1984. Soil nutrient bioavailability. John Wiley & Sons, Inc., New York.

Fujii, R., L.L. Hendrickson, and R.B. Corey. 1983. Ionic activities of trace metals in sludge-amended soils. Sci. Total Environ. 28:179–190.

Nye, P.H., and P.B. Tinker. 1977. Solute movement in the soil-root system. University of California Press, Berkeley, P. 42.

Soil Science Society of America. 1987. Glossary of soil science terms. Soil Science Society of America, Madison, WI.

Wietholter, S. 1983. Predicting potassium uptake by corn in the field using the strontium nitrate soil testing method and a diffusion-controlled uptake model. Ph.D. thesis. Univ. of Wisconsin-Madison (Diss. Abstr. 83-21780).

3 Soil Test Calibration[1]

Clyde E. Evans[2]

The purpose of soil testing is to gain information about a soil and to provide a fertilizer recommendation that is economically sound. It can be considered as a way of transferring research information to soils on which field experimental data are not available. For example, the most reliable procedure of obtaining an accurate fertilizer recommendation would be to conduct a fertility trial on each field for each crop. Realizing that extensive research such as this is not practical, the most satisfactory way to transfer soil testing technology is by arranging crops and soils into manageable groups.

CALIBRATION IS BASIC

Soil test calibration is basic to a good soil testing program. Without the background of information on crop response as related to soil levels of plant nutrients, the values for extracted nutrients have little meaning. Several extracting solutions have been developed that satisfactorily extract nutrients from the soil. The precision in chemical procedures far exceeds that of other steps such as sampling and interpretation.

Calibration is rather closely related to correlation and interpretation. Since correlation and interpretation are covered in other chapters in this publication, only brief mention of them is needed here. This chapter will primarily be directed toward concepts and recent information on calibration, and toward pointing out continuing research needs to maintain current information.

A properly calibrated soil test should provide information in two categories. It should:

1. Identify the degree of deficiency or sufficiency of an element.
2. Identify how much of the element should be applied if it is deficient.

Both of these must be considered in collecting data for soil test calibration. Information identifying the degree of deficiency or sufficiency can be obtained without answering the question of how much of the element is needed

---

[1] Contribution from the Alabama Agricultural Experiment Station, Auburn University, Auburn, AL 36849.

[2] Professor of Soil Fertility, Department of Agronomy and Soils, Auburn University, Auburn, AL 36849.

for optimizing yield. This involves comparison of a zero or check treatment yield with that of a treatment in which all other nutrients are at an adequate level. In conducting field experiments, information covering both categories can be obtained simultaneously. With all nutrients except the one being observed held at an optimum level, rates of the nutrient under observation can be applied and yield taken. In order to define the response curve it is essential to include rates of the nutrient above and below the point of maximum yield. The check treatment gives the information on yield without the element, and all other rates will help identify the point of maximum yield. For practical considerations, economics must be taken into account and this can require careful observation as the yield response approaches maximum.

FIELD EXPERIMENTS

When collecting soil test calibration data, it is necessary to have values covering a range from deficient to sufficient in order to establish a relative yield curve. There should be enough values to provide good distribution over the full range of the response curve. This can be accomplished by using several locations with possibly only one replication per location; or by selecting a soil that is deficient and establishing levels, and then superimposing rates of fertilizer on the different levels at one location. Research at Auburn University has included both procedures.

Jordan et al. (1966) conducted experiments on Coastal bermudagrass [*Cynodon dactylon* (L.) Pers.] at several sites in a local area. These were nonreplicated experiments on soils with similar characteristics and included a broad range of soil test values for phosphorus (P) and potassium (K). The

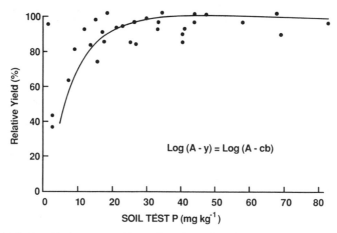

Fig. 3-1. Relationship between soil test P and relative yield of Coastal bermudagrass. (Jordan et al., 1966).

data from these experiments for P and K using a modification of the Mitscherlich equation as proposed by Bray (1948) are given in Fig. 3–1 and 3–2, respectively. Figure 3–1 indicates maximum yield at about 30 mg P/kg extracted with the Mehlich I (double acid) extractant (Mehlich, 1953). The yield curve declined rapidly with decreasing soil test values <20 mg P/kg. For K in Fig. 3–2, maximum yield occurred at about 40 mg K/kg. Two response curves were calculated for the K relationship. Because the equation approaches a soil test value of zero at zero yield, the dotted line was calculated using a correction of 10 mg/kg for all soil test values. Since soil test values for K do not go to zero, as is sometimes the case with P, the corrected line appears to better represent the data in the low yield range.

In the above experiment, the relationship between relative yield and soil test value was established. Although there were two rates of P and K, the response to optimum rate of fertilizer was not identified. In a later experiment, a soil low in both P and K was identified and levels of these elements were established on blocks. In addition to the "low" blocks, corrective applications of P and K were applied to give "medium" and "high" blocks. Rates of P and K were then applied on each of the blocks. With 224 kg N/ha, 90 kg K/ha on low fertility, and 45 kg K/ha on medium fertility levels gave near maximum yield without soil K depletion.

Three-year average yields for this experiment are given in Table 3–1. During the 3-year period, soil K was depleted in the check treatment of the high fertility blocks so that yields were reduced at zero K in the second and third years. This indicated that soil K can be rapidly depleted by forage removal from a coarse-textured soil.

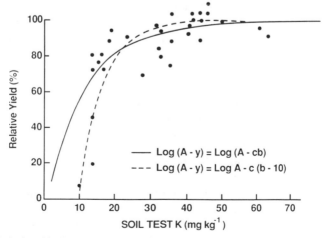

Fig. 3–2. Relationship between soil test K and relative yield of Coastal bermudagrass. (Jordan et al., 1966).

Table 3-1. Yields of Coastal bermudagrass forage at three fertility levels with varying K rates at 224 kg N/ha.

K rate	Year		
	1965	1966	1967
kg/ha	relative yield %		
	Low fertility level		
0	75	64	59
45	85	86	87
90	90	93	95
180	87	98	94
360	95	99	97
	Medium fertility level		
0	91	82	73
45	99	83	85
90	95	88	93
180	99	97	94
360	99	103	93
	High fertility level		
0	91	79	78
45	95	82	90
90	95	93	92
180	98	96	93
360	100	100	100
	100% yield, kg/ha		
	13 814	10 926	9 376

USE OF RELATIVE YIELDS

Relative yields can be used effectively in summarizing calibration data. The objective of calibration is to develop yield response equations over all sites for similar soils so that appropriate inputs of fertilizers can be made to obtain economically optimum yields. It is recognized that local soil and climatic conditions existing at each site may have measurable effects on productivity.

A wide scattering of absoute yields may occur as a result of factors other than soil fertility. This scattering of absolute yields does not necessarily mean there is poor correlation, but there may be a better relationship by using relative yields to eliminate some of the site influences. Examples of the use of relative yield vs. absolute yield are given in Fig. 3-3 and 3-4. Although this only takes into account the differences in 2 years during which there is no large difference in absolute yields, this concept is demonstrated by the figures. One could readily speculate how the scatter for absolute yields might be considerable if additional years and site differences were involved.

Relative yields are calculated according to the following relationship:

$$\text{Relative yield (\%)} = \frac{\text{Yield from the test plot}}{\text{Yield from fully treated plot}} \times 100. \quad [1]$$

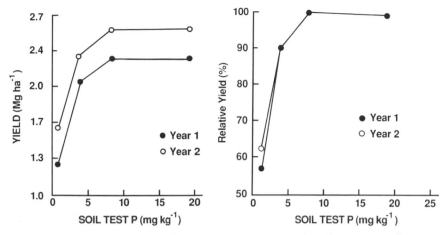

Fig. 3–3. Relationship between soil test P and yield of soybean.

Fig. 3–4. Relationship between soil test P and relative yield of soybean.

A satisfactory calibration should have all the fertilizer elements at optimum except the one being studied. In calibrating for K in an N, P, K factorial experiment, the following relationship can be used:

$$\text{Relative yield (\%)} = \frac{\text{Yield from test plot with NP}}{\text{Yield from plot with NPK}} \times 100. \qquad [2]$$

Although the relative yield concept has its advantages, in that it usually provides better calculations for summarizing calibration data, the shortcoming of its economic consideration cannot be overcome even by the improved fit of plotted data.

CALIBRATION IS A CONTINUING PROCESS

There is a need to continually verify and update soil test calibration information. Perhaps many laboratories began operation with limited calibration information. As more attention is directed toward soil test research, needed changes should be incorporated.

Changing management practices also bring about the necessity for continually conducting soil test calibration research. In Alabama, soil test summaries in recent years have shown decreasing P levels in soils in some areas of the state. The two areas with the greatest decrease are the most heavily row-cropped land. The reason for the declining P fertility levels is not clear. One possibility is that deeper tillage is diluting the P with a larger volume of subsoil that is more acid, has a higher clay content, and a greater P fixing capacity. Another is that erosion may be removing topsoil rich in P. A third possibility is that growers are using less fertilizer than recommended. The decline in residual soil test P has not been accompanied by a decrease in yields in recent years; actually the reverse is true.

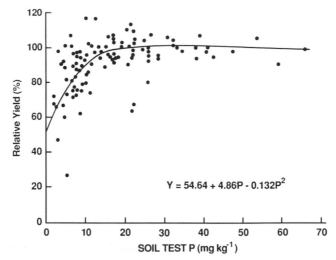

Fig. 3-5. Relationship between soil test P and relative yield of seven crops. (Equation applies to the response portion of the curve below 100%.)

The decline in P levels from farmers' samples does not appear to be consistent with fertilizer recommendations. It has been shown repeatedly that relatively low rates of P fertilizer will increase soil test P levels. Early calibration research in Alabama has demonstrated that rates of P fertilizer recommended for soils low or medium in P are adequate to give maximum economic yield and to raise the level of extractible P (Rouse, 1968; Cope, 1984).

Auburn University is currently carrying out extensive on-the-farm research in the two areas mentioned above. A recent summary of results indicates that the P ratings need to be adjusted to fit the current calibration. The data for seven crops on soil groups 1 and 2 are given in Fig. 3-5. The maximum yield occurred at about 18 mg P/kg and yields were 75% of maximum at about 5 mg P/kg of extractible P. The current values used in interpretation for these levels are 25 and 12.5 mg P/kg.

Other research has also indicated that soil test ratings should possibly be revaluated. Experiments in Kentucky led to the conclusion that the predicted values for near maximum yield of corn (*Zea mays* L.) and soybean [*Glycine max* (L.) Merr.] are below those currently being used for fertilizer recommendations (Thom, 1985). In an experiment on sugar cane (*Saccharum officinarum* L.), Thomas et al. (1985) found no response to P fertilizer on a soil considered low (4.9 mg $NaHCO_3$–extractable P/kg) in P and had P leaf concentrations below the considered optimum level. There was no correlation of yield with $NaHCO_3$-extractable P but there was a good correlation between leaf P and residual soil P ($r = 0.83$). This indicates that there was a relationship between the soil extracted P and plant uptake but that the lowest residual P level by this extraction procedure was adequate for sugar cane growth.

SUMMARY

1. Soil testing is a reliable procedure for transfering research to growers' soils. The usefulness of the information is dependent upon good calibration information.
2. Field experiments for calibration should cover a broad range of conditions and are needed for each geographic area in which recommendations are made.
3. The use of relative yield permits the combining of data from many locations or for many years into a single plot.
4. Calibration should be a continuous process in order that the latest research information is incorporated into the soil test program, so that soil test interpretation and fertilizer recommendations are based on current conditions.

REFERENCES

Bray, R.H. 1948. Correlation of soil test with crop response to added fertilizers and with fertilizer requirement. p. 53–85. *In* H.B. Kitchen (ed.) Diagnostic techniques for soils and crops. The American Potash Institute, Washington, DC.

Cope, J.T. 1984. Long term fertility experiments on cotton, corn, sorghum and peanuts 1929–82. Alabama Agric. Exp. Stn. Bull. 561.

Jordan, C.W., C.E. Evans, and R.D. Rouse. 1966. Coastal bermudagrass response to applications of P and K as related to P and K levels in the soil. Soil Sci. Soc. Am. Proc. 30:477–480.

Mehlich, A. 1953. Determination of P, Ca, Mg, K, Na, and NH_4. North Carolina Dep. of Agric. Soil Test Div. Mimeo, 1953.

Rouse, R.D. 1968. Soil test theory and calibration for cotton, corn, soybeans, and coastal bermudagrass. Alabama Agric. Exp. Stn. Bull. 375.

Thom, W.O. 1985. Soil test interpretation with corn and soybeans on a belknap Silt Loam. Kentucky Agric. Exp. Stn. Bull. 720.

Thomas, J.R., A.W. Scott, Jr., and R.P. Wiedenfeld. 1985. Fertilizer requirements of sugarcane in Texas. J. Am. Soc. Sugar Cane Technol. 4:62–72.

4

Role of Response Surfaces in Soil Test Calibration[1]

Larry A. Nelson[2]

There are several stages in the process of calibrating a soil test procedure. One occurs in the laboratory where adsorption isotherms are obtained for specific soils using the soil test procedure. Based upon these laboratory results, preliminary studies are then conducted in the greenhouse with crops such as corn (*Zea mays* L.) and millet (*Setaria italica*) to determine the rates of fertilizer required to obtain optimal fertilizer response. A third phase is conducted in the field to see if the optimal rates obtained in the greenhouse should be adjusted for field conditions. Although response surfaces could theoretically be used in the greenhouse, it is the field phase of research where response surfaces become most useful.

It is usually necessary to conduct a series of experiments in a network of sites in order to calibrate soil tests. Similar experimental and treatment design throughout the series is necessary, as is similar management of the experiments. A paper describing several closely related methods of calibrating soil test data with field response for such a series of sites was presented approximately 10 years ago by Nelson and Anderson (1977) at the 1975 ASA meetings in Knoxville, TN. Their approach was to relate delta yield to the soil test determinations for a series of sites. *Delta yield* was defined as the difference between the yield of the treatment having no fertilizer and the yield of the treatment with all factors at adequate, but not excessive, levels. The advantage of working with delta yields rather than actual yields was that much of the among-site variation in levels of actual yields was due to factors other than fertilizer, such as climate and permanent soil properties. Two points on the response curve defined the delta yield, which was the only use of the response curve. Emphasis was placed more on the soil test data and their interpretation than upon the yield response curve. One soil nutrient was calibrated at a time, although multinutrient calibration was discussed briefly.

The approach in this chapter is to describe the response to added fertilizer by fitting a response surface mathematically to the yields obtained from the single experiment or to the yields obtained from all of the sites of the

---

[1] This work was sponsored by the Department of Statistics, North Carolina State University, Raleigh,

[2] Professor of Statistics, North Carolina University, Raleigh, NC 27695.

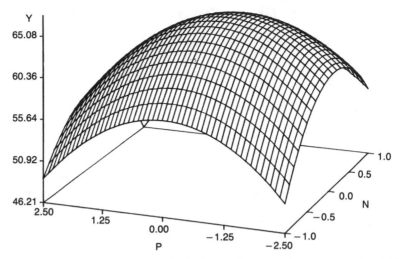

Fig. 4-1. Yield response surface for a single site as a function of two controlled variables.

series. Points on the response surface are average yields that were obtained from fitting a regression equation of the experimental yield data on the controlled rate combinations of two or more added fertilizer nutrients. The entire surface is smooth because it is generated by a mathematical model such as a second-order polynomial. An example of a response surface in two controlled variables for a single site is shown in Fig. 4-1. Because yields contain a contribution from both the soil nutrient and the added nutrient, our mathematical model will necessarily involve variables due to the soil nutrient (called covariables) as well as that to the added nutrient. The inclusion of covariables in the model may not be necessary for modeling the within single site variation, especially if there is very little range in values of the covariable within that site.

Ultimately one would want to develop a more general model over a series of sites to benefit from the among-site range in soil nutrient levels in estimating the coefficients for the soil test variables. In many past studies, emphasis was placed on the response surface of the variables for added nutrients and the soil nutrient variables were treated as nuisance parameters. The emphasis is currently placed on the soil test rather than on the response to the added nutrient. Because we are not dealing with delta yields as we were 10 years ago, it is necessary to contend with the problems of variation due to permanent soil properties and climatic variables because they are responsible for much of the variation among sites. One difficulty with the permanent soil properties is to find those that apply to a given set of data. Sorting out and modeling the climatic variables is a much greater problem. Unless variation due to these two sources is included in the model, the error estimate for the regression will be unrealistically high and it will be difficult to make sensitive tests of significance for the soil test and added nutrient variables. In addition to the error estimate problem, one has to consider the effect of the permanent soil properties and climatic factors and how they are modeled on the estimates of the regression coefficients for the soil nutrient.

Despite some of the modeling considerations mentioned above, a number of workers have used the covariance approach (Voss & Pesek, 1967, Voss et al., 1970; Ryan & Perrin, 1973; Colwell, 1967, 1968, 1970, 1979). It has the advantage of ease of fitting by least squares techniques, even for the multinutrient case. The emphasis of these researchers focused on the adjustment of a response surface in the added nutrient for the soil nutrient level rather than an attempt to calibrate soil tests.

The mathematical form for the covariance model for a single nutrient is as follows:

$$\hat{Y} = \beta_0 + \beta_1 n + \beta_2 d + \beta_3 n^2 + \beta_4 d^2 + \beta_5 nd \qquad [1]$$

where $\hat{Y}$ = a predicted yield on the response curve; β_0 = intercept; $\beta_1, \ldots, \beta_5$ = regression coefficients due to added and soil nutrients; n = applied nutrient; and d = soil nutrient.

It should be noted that the covariance model is flexible; for example, if the relationship between yield and d is linear, the quadratic term may be omitted. Also, the interaction term between n and d could be omitted if there is no indication of interaction. It is realistic to assume that there should be other terms in Eq. [1] to explain variation among sites due to permanent soil properties and climatic factors.

SELECTION OF SOIL TEST PROCEDURE

It is reasoned that part of the process of calibrating soil tests is the selection of a good soil test procedure that works well for the particular soil being tested. Thus, several versions of Eq. [1], which differ only with respect to soil test procedure, may be used. The procedure whose regression gives the highest R^2 value would then be selected for making soil tests for that nutrient in the future, other factors being equal. Another aspect of the process is that once the best soil test procedure has been selected it is necessary to determine the relationship between the soil test readings and yield when other variables in the model are held constant. If the relationship between the soil test nutrient and yield is linear, the regression coefficient for the soil test nutrient gives a direct estimate of the yield response per unit change in soil test reading. If the relationship is curvilinear, the change in yield per unit change in soil test reading will vary depending upon the range in which the soil test variable is being considered.

Table 4-1. Format of analysis of variance table for the covariance model
(two nutrients).

Source	df	
Treatment (T)	$t - 1$	
N		1
P		1
N^2		1
P^2		1
NP		1
Lack of fit		$t - 6$
Location (L)	$l - 1$	
Soil N		1
Soil P		1
(Soil N)^2		1
(Soil P)^2		1
(Soil N) $\times$ (Soil P)		1
Other site variables	$l - 6$	
(such as permanent soil properties and climatic variables)		
Replications in location	$(r - 1) l$	
$T \times L$	$(t - 1)(l - 1)$	
25 single degree of freedom interactions		1
		.
		.
		.
		.
		.
		etc.
Lack of fit	$(t - 1)(l - 1) - 25$	
Error	$l (t - 1)(r - 1)$	

DETERMINING RELATIVE IMPORTANCE OF
SOIL AND ADDED NUTRIENT

It may be desirable to compare the regression coefficient due to the added nutrient with that due to the soil test unit. In order to do this one should convert the two regression coefficients to the same scale. A possible method of interpretation is to make a test of significance to determine if the two coefficients are the same. One could also test the hypothesis, for example, that $b_{\text{soil}} = 1.5 \ b_{\text{added nutrient}}$ with a simple t test.

Another useful interpretational device is to obtain the analysis of variance for the covariance with a decomposition of the regression sum of squares into individual degree of freedom components. An example is presented in Table 4-1. The sums of squares for the individual degrees of freedom for the added nutrients are subdivisions of the treatment sum of squares. The sums of squares for the individual degrees of freedom for the site variables (covariables) are subdivisions of the site sum of squares. The individual degrees of freedom for the interaction between site variables and added nutrient model terms are subdivisions of the sum of squares for the interaction between sites and fertilizer treatments.

In the analysis of variance for the regression shown in Table 4-1, the linear effect for the added nutrients should be first in order, the quadratic

Table 4-2. Key-out of sources of variation, degrees of freedom, and expected mean square for a combined analysis of variance (Laird & Cady, 1969).

Source of variation	df	Expected mean squares
Sites (S)	$(s - 1)$	
Blocks/sites	$s (b - 1)$	$\sigma_1^2 + t \, \sigma_2^2$
Treatment (T)	$(t - 1)$	
$S \times T$	$(s - 1) (t - 1)$	
Combined experimental error	$s (b - 1) (t - 1)$	σ_1^2

effects should be second, and the mixed terms for the same form of nutrient, third. The linear effect due to the soil nutrients should follow, and curvilinear and mixed terms for the soil nutrients placed next if they occur in the model. Type I sums of squares in the interpretation might be used rather than Type II or Type III. This allows for each variable adjusted for all other variables preceding it in the model. The soil nutrient variables are adjusted for the entire group of added nutrient variables. If there is more than one soil nutrient, a regression model should be run for each possible order of the soil nutrients. Laird and Cady (1969) presented the appropriate analysis of variance for data combined over locations and showed which terms should be considered as error terms for the various site, reps(site), treatment, and treatment × site variables for a given assumed model. Their model is shown in Table 4-2. Pantula et al. (1985) confirmed that such an analysis of variance model was appropriate for estimating the error terms unless the covariables (site variables) were measured by plot rather than by site. They presented a general least squares procedure that could be used for fitting the response surface and covariables to combined data from the sites (or from sites and years) regardless of whether the covariables were read by plot or by site.

HILDRETH MODEL APPROACH

An alternative model approach was given by Hildreth (1957) in which soil nutrient and added nutrient are combined into a single variable. The model for such a form for a single nutrient follows:

$$\hat{Y} = \alpha_1 X + \alpha_2 X^2 \qquad [2]$$

where $\hat{Y}$ = predicted yield; α_1, and α_2 = regression coefficients; X = total nutrient $(n + \lambda d)$; where n = added nutrient; d = soil nutrient; and λ = unknown factor of proportionality. The λ may be interpreted as the ratio between the marginal response to soil nutrient and the marginal response to added nutrient for any given level of total nutrient.

In Eq. [2] it is assumed that yield is zero when the total amount of nutrient is zero. By substituting $(n + \lambda d)$ for X in Eq. [2], the covariance model that is being called Eq. [3] is obtained,

$$\hat{Y} = \alpha_1 (n + \lambda d) + \alpha_2 (n + \lambda d)^2 \qquad [3]$$

where all terms are defined as before.

By expanding and collecting terms, Eq. [4] is obtained as

$$\hat{Y} = \beta_1 n + \beta_2 d + \beta_3 n^2 + \beta_4 d^2 + \beta_5 nd \qquad [4]$$

where

$$\beta_1 = \alpha_1; \ \beta_2 = \alpha_1\lambda; \ \beta_3 = \alpha_2; \ \beta_4 = \alpha_2\lambda^2; \ \beta_5 = 2\alpha_2\lambda.$$

It should be pointed out that Eq. [1] and [4] are identical except for the intercept in Eq. [1]. Equation [1], however, assumes separate relationships between $\hat{Y}$ and soil nutrient and $\hat{Y}$ and added nutrient. Therefore Eq. [1] is more flexible with respect to the inclusion or exclusion of terms in or from the model than is Eq. [4]. Also, the calculation of d has not yet been discussed.

Hildreth (1957) pointed out that estimating the parameters of Eq. [4] is an ordinary least squares problem except for some restrictions that must be imposed in the fitting, i.e.,

$$(\beta_2/\beta_1) = (\beta_4/\beta_3)^{1/2} = \beta_5/2\beta_3 = \lambda.$$

Hildreth discussed nonlinear methods for fitting subject to these restrictions at length. A more efficient way of estimating λ is to fit Eq. [3] directly, using nonlinear least squares. In this instance there would be fewer coefficients to estimate, λ would be estimated directly, and a standard error for λ would be available.

Hurst and Mason (1957) and later Anderson and Nelson (1971) used the Hildreth model, but to simplify the fit they assumed that $\lambda = 1$. This results in soil nutrient being expressed in units that are equivalent to added nutrient. Substituting $(n + d)$ for X in Eq. [2] and omitting the intercept, the model then becomes

$$\hat{Y} = \alpha_1 (n + d) + \alpha_2 (n + d)^2, \qquad [5]$$

where all term are defined as before.

Model [5] has an advantage over the more general covariance approach because it should produce a more efficient estimate of nutrient effect if the total nutrient involves only one term rather than if it is split into two parts. The total nutrient will have a wider range than either of the component parts, which results in a more efficient estimate of the regression coefficient for the nutrient.

One of the goals of both teams of researchers in using Eq. [5] was to estimate the amount of nutrient in the soil, d, in equivalent units of added nutrient. In so doing, they assumed that d did not vary greatly within the site, and therefore an average d could be estimated for the site rather than assuming that soil nutrient should be modeled by plot as Hildreth did.

Anderson and Nelson (1971) noted that the amount of estimated d varied greatly with the form of the mathematical model used. In cases where the quadratic coefficient was quite small, the estimates could increase significantly due to division by a small number in the formula for estimating d. They found cases where the quadratic model gave a highly biased estimate of d that was exaggerated when a multinutrient model in N, P, and K was used and a simultaneous estimate of soil levels of all these nutrients was made.

Mombiela et al. (1981) regressed d estimates obtained by the Hurst and Mason (1957) and Anderson and Nelson (1971) methods on functions of soil test readings from a set of fertilizer phosphorus (P)-potato (*Solanum tuberosum*) yield data from Maine and a corresponding set of data from North Carolina. They found the relationship between d and the soil test reading to be linear and strong ($\hat{d} = c_1 t$, where c_1 is a linear proportionality constant and t is the soil test reading). They also compared this calibration procedure using d estimates obtained with three different yield response models, the Mitsherlich, the quadratic, and square root model. [See Anderson & Nelson (1971) for a description of these models.] These researchers found that for the particular sets of data studied, the Mitscherlich and square root models provided a better fit than did the quadratic model, which was unable to accomodate to the initial high crop yield response to total P and subsequent flattening of the curve. They concluded that the estimate of d is a stable measurement that appears to be better than yield for use in soil test calibration purposes. It would seem that d provides an estimate of the amount of nutrient in the soil (expressed in units of freshly added nutrient) per unit soil test.

As mentioned earlier, the d's are difficult to estimate in the multinutrient case, and estimates of d can be biased. For this reason, the Mombiela et al. (1981) procedure may not be workable in the multinutrient case. Further work along these lines is needed.

As an extension of the Mombiela et al. (1981) method, replacing d in Eq. [5] by $c_1 t$ with λ assumed equal to 1, should be considered, because t is known but c_1 must be estimated from the data (along with the regression coefficients). Upon expansion, Eq. [6] is obtained,

$$\hat{Y} = \alpha_1 (n + c_1 t) + \alpha_2 (n^2 + 2nc_1 t + c_1^2 t^2). \qquad [6]$$

Equation [7] is obtained upon collecting terms

$$\hat{Y} = \beta_1 n + \beta_2 t + \beta_3 n^2 + \beta_4 t^2 + \beta_5 nt \qquad [7]$$

where

$$\beta_1 = \alpha_1; \ \beta_2 = \alpha_1 c_1; \ \beta_3 = \alpha_2; \ \beta_4 = \alpha_2 c_1^2; \ \beta_5 = 2\alpha_2 c_1.$$

Again the following restrictions must be imposed on the fitting.

$$(\beta_2/\beta_1) = (\beta_4/\beta_3)^{1/2} = \beta_5/2\beta_3 = c_1.$$

Fitting can be achieved in the same manner as in the Hildreth model. It is clear that the above is just another form of the Hildreth equation with c_1 representing λ and t representing an estimate of d. However, this relationship exists only if it is assumed that the intercept relating d to t is zero and the model is linear. If the model is curvilinear, the Mombiela et al. approach will produce a model different than that of Hildreth.

Again, fitting Eq. [6] directly using nonlinear least squares would be preferable to fitting Eq. [7] because fewer coefficients need to be estimated. The value of c_1 is estimated directly and its standard error is available when Eq. [6] is fitted.

The Mombiela et al. approach (1981) should best be used to identify the form of the model for each nutrient separately, then the appropriate terms should be placed in the full multinutrient model and their parameters estimated with the remaining parameters of the multinutrient model. Mombiela et al. (1981) found that the relationship between d and soil test reading is linear and that the lines pass through the origin for two sets of potato data. This makes the fitting more straightforward but seems to be an assumption that is too restrictive. It is more realistic to assume that the form of the relationship would vary by soil, crop, form of the model for response to added nutrient, and soil test procedure. Perhaps one criterion that should be used for selecting a good soil test procedure would be that it provide a linear relationship with d.

There is a great deal of research needed to study the effect of biases in the estimate of the d's due to the use of an inappropriate model for the estimate of c_1 for each nutrient. It is not clear that fertilizer response models that produce biased estimates of d will cause the same degree of problems with estimates of c_1. Researchers also need to study the effect of how soil properties, other than soil nutrients under investigation and the climatic variables, affect the estimates of c_1 for the various nutrients.

Equations [6] and [7] appear to have a considerable advantage over Eq. [5] in obtaining the equivalence of soil test reading to added fertilizer equivalent. They provide a more direct route than that available from estimating d and then regressing it upon the soil test readings. The approach of Eq. [6] and [7] is quite general in the sense that it fits in well with a number of different forms of models for the response to added nutrient. For example, Mombiela et al. (1981) chose as their final scheme a Mitscherlich model in which soil test P was included in the exponent. The fitting was done by nonlinear least squares with an estimate of the c_1 value for soil test P being one of the results of the fitting. This modified Mitscherlich model was used in preference to the quadratic equation because as mentioned previously, the quadratic did not fit the data well.

An extension of the procedure described above as well as in Eq. [6] and [7] to the multinutrient case is obvious. For two nutrients, Eq. [8] is obtained

$$\hat{Y} = \beta_1 n + \beta_2 t_1 + \beta_3 n^2 + \beta_4 nt_1 + \beta_5 t_1^2 + \beta_6 p + \beta_7 t_2 + \beta_8 p^2$$
$$+ \beta_9 pt_2 + \beta_{10} t_2^2 + \beta_{11} np + \beta_{12} nt_2 + \beta_{13} pt_1 + \beta_{14} t_1 t_2 \qquad [8]$$

where n is the first added nutrient, p is the second added nutrient, t_1 is the soil test reading for the first soil nutrient, and t_2 is the soil test reading for the second soil nutrient.

The following restrictions must be imposed when fitting the regression in the two-nutrient case

$$(\beta_2/\beta_1) = \beta_4/2\beta_3 = (\beta_5/\beta_3)^{1/2} = \beta_{14}/\beta_{12} = c_1$$

$$(\beta_7/\beta_6) = \beta_9/2\beta_8 = (\beta_{10}/\beta_8)^{1/2} = \beta_{14}/\beta_{13} = c_2.$$

Again, fitting may be achieved more directly using nonlinear least squares to estimate c_1 and c_2 as terms in the model as described previously. However, in the form presented above, Eq. [8] does not include terms for permanent soil properties and climatic variables. These should be included and this complicates the whole process. Considerable work needs to be done to evaluate the interplay between permanent soil properties and climatic variables with the c_1 estimation process. The procedure also needs a great deal of fine tuning by the use of a variety of data sets. It appears that estimation of the model parameters is not difficult with modern computing equipment, but the modeling itself is more difficult.

SUMMARY

Statistical procedures for calibrating soil tests from response surface fitting of response to added fertilizer are discussed. Estimation of the added fertilizer equivalent of soil nutrient as measured by soil test rather than estimation of the response surface per se is emphasized. The major advantage of these procedures over the traditional soil test calibration statistical procedures is the possibility of simultaneously calibrating soil tests for several soil nutrients. Estimates may be obtained directly for both the added fertilizer–soil test equivalent and its standard error for each nutrient. The resulting estimates appear to be very useful to those responsible for choosing soil test procedures and those responsible for interpreting soil test results. The methods described show considerable promise over some of the traditional procedures for soil test calibration.

The general covariance model, which includes terms for added nutrient, soil test results, permanent soil property variables, and climatic variables is discussed. The Hildreth model, including recent modifications is also discussed. Methods to determine the relationship between estimated soil nutrient (d) and soil test are described, but it is pointed out that direct estimation of the added fertilizer–soil test equivalent is preferable to a two-stage estimation procedure involving an initial estimation of d. A complicating factor in the modeling process is the necessary inclusion of permanent soil properties and climatic variables in the model. There is some question as to how these variables should be modeled. There needs to be a great deal of fine tuning of the method using a variety of data sets. Statistical fitting procedures are discussed briefly.

REFERENCES

Anderson, R.L., and L.A. Nelson. 1971. Some problems in the estimation of single nutrient functions. p. 203–222. *In* Bulletin of the International Statistical Institute, 44, Part 1. International Statistical Institute, The Hague, Netherlands.

Colwell, J.D. 1967. Calibration and assessment of soil test for estimating fertilizer requirements. I. Statistical models and tests of significance. Aust. J. Soil Res. 5:275–293.

Colwell, J.D. 1968. Calibration and assessment of soil test for estimating fertilizer requirements. II. Fertilizer requirements and an evaluation of soil testing. Aust. J. Soil Res. 6:93–103.

Colwell, J.D. 1970. A comparison of soil test calibrations for the estimation of phosphorus fertilizer requirements. Aust. J. Exp. Agric. Anim. Husb. 10:774–782.

Colwell, J.D. 1979. National soil fertility project, Vol. 2. Commonwealth Scientific and Industrial Research Organization (CSIRO), Canberra, ACT, Australia.

Hildreth, C.G. 1957. Possible models for agroeconomic research. p. 176–186. *In* E.L. Baum et al. (ed.) Economic and technical analysis of fertilizer innovations and resource use. Iowa State University Press, Ames.

Hurst, D.C., and D.D. Mason. 1957. Some statistical aspects of the TVA-North Carolina cooperative project on the determination of yield response surfaces for corn. p. 207–216. *In* E.L. Baum et al. (ed.) Economic and technical analysis of fertilizer innovations and resource use. Iowa State University Press, Ames.

Laird, R.J., and F.B. Cady. 1969. Combined analysis of yield data from fertilizer experiments. Agron. J. 61:829–834.

Mombiela, F.A., J.J. Nicholaides, III, and L.A. Nelson. 1981. Method to determine the appropriate mathematical form for incorporating soil test levels into fertlizier response models for recommendation purposes. Agron. J. 73:937–941.

Nelson, L.A., and R.L. Anderson. 1977. Partitioning of soil test—crop response probability. p. 19–37. *In* M. Stelly (ed.) Soil testing: Correlating and interpreting the analytical results. Special Pub. 29. American Society of Agronomy, Crop Science Society of America, and Soil Science Society of America, Madison, WI.

Pantula, S.G., L.A. Nelson, and R.L. Anderson. 1985. Estimation of linear models for field experiments. Common. Stat.-Theory Methods 14:2199–2217.

Ryan, J.G., and R.K. Perrin. 1973. The estimation and use of a generalized response function for potatoes in the Sierra of Peru. North Carolina Agric. Exp. Stn., Tech. Bull. no. 214.

Voss, R., J.J. Hanway, and W.A. Fuller. 1970. Influence of soil management and climatic factors on the yield response by corn (*Zea mays* L.) to N, P, K fertilizer. Agron. J. 62:736–740.

Voss, R., and J.T. Pesek. 1967. Yield of corn as affected by fertilizer rates and environment factors. Agron. J. 59:567–572.

5 Soil Testing Interpretations: Sufficiency vs. Build-up and Maintenance[1]

R. A. Olson, F. N. Anderson, K. D. Frank,
P. H. Grabouski, G. W. Rehm,
and C. A. Shapiro[2]

Very little research on calibration of soil testing had been done prior to the 1950s and little faith existed among most professional agronomists except for its use in the diagnosis of acid and alkali conditions. During the intervening years, most of the background research validating soil testing as a means for predicting crop nutrient needs has been equated with the sufficiency concept for interpreting test results. Primary ranges were established of low, medium, high with respective interpretations of crop response to applied nutrient being assured, possible, and unlikely, accordingly, a concept of "fertilizing the crop."

The above-noted historical approach to soil test interpretation has been modified toward the build-up and maintenance concept by many laboratories, one of "fertilizing the soil," contributing to the vast increase that has occurred in fertilizer consumption in the past 30 years. This approach affords a much more liberal fertilizer recommendation because there is no real cut-off level—even with a high test, nutrients are recommended to replace the amount likely to be removed by the crop to be grown. The native capacity of a majority of arable soils to support modest crop yields from native mineral/organic nutrient reserves is thereby discounted. Furthermore, in order to be completely honest, the use of this concept should advocate replacing (maintaining) all of the 13 soil-derived nutrients that the crop is projected to remove and not just nitrogen (N), phosphorus (P), and potassium (K).

Clearly there are vested interests that prefer the build-up and maintenance approach over the sufficiency concept in soil test interpretation; the more fertilizer sold the better the immediate business of the fertilizer dealer and his supplier. This has led to the impasse that presently exists between some commercial enterprises and university agronomists on soil test interpretation. There are, of course, major inherent soil differences that exist locally and regionally from pedogenic processes that impact on the situation. Soils of

[1] Contribution of the Nebraska Agricultural Experiment Station, Lincoln, NE.

[2] Professor, and Extension Soils Specialists, Department of Agronomy, University of Nebraska, Lincoln, NE (Rehm now at the University of Minnesota, St. Paul, MN).

Table 5-1. Chemical characteristics of a representative soil in the Mollisol order (midwestern USA) compared with one of the Ultisol order (southeastern USA).

	Soil	
Soil characteristic	Sharpsburg silty clay loam (Nebraska)	Georgeville silty clay loam (North Carolina)[†]
Taxonomic classification	Typic Argiudoll	Typic Hapludult
Dominant clay mineral	Montmorillonite	Kaolinite
Cation exchange capacity, $cmol_c/kg$	23.8	5.67
Exchangeable K^+, $cmol_c/kg$	0.9	0.25
Exchangeable Mg^{2+}, $cmol_c/kg$	3.6	1.44
Exchangeable Ca^{2+}, $cmol_c/kg$	15.3	3.91
Exchangeable Al^{3+}, $cmol_c/kg$	0	0.05
Organic matter, g/kg	32	22
Bray and Kurtz Pl, mg/kg	20	4
pH (H_2O)	6.4	5.5

[†] Information supplied by E.J. Kamprath, 1985, personal communication.

the warm and humid subtropics and tropics having kaolinitic and sesquioxide clays, low organic matter, high fixing capacity for some elements, and resulting low inherent nutrient delivery capacities are more likely candidates for the build-up/maintenance concept than soils of the Corn Belt having 2:1 type clays and higher organic matter contents. Table 5-1, depicting chemical characteristics of a representative soil in the Mollisol order of the midwestern USA in contrast with those of an Ultisol in the southeastern part of the USA, exemplifies the inherent regional differences that exist. The disparity depicted in this table does not take into account the large differences that exist in subsoil nutrient reserves as well.

Not to be overlooked in the argument over the two concepts is the fact that the introduction of inorganic fertilizer has done more to enhance the agricultural productivity of this country since World War II than any other factor. A reasonable estimate would suggest that at least 50% of all grain crop production is directly attributable to applied fertilizer. Care must be exercised so that contentions against excessive fertilizer use are not taken out of context and used to discourage inorganic fertilizer use per se.

THE NEBRASKA SOIL TEST COMPARISON STUDY

A long-term soil test comparison study was undertaken in Nebraska beginning in 1973 (Olson et al., 1982a). The study was prompted by the wide discrepancy in fertilizer recommendations made to farmers from the four major commercial laboratories that handled approximately 80% of the soil testing business in the state at the time. All of the commercial laboratories were apparently using build-up/maintenance approach, compared with the University's Extension Service Laboratory's sufficiency approach. The experiment was conducted on four major soil series located on four of the state's

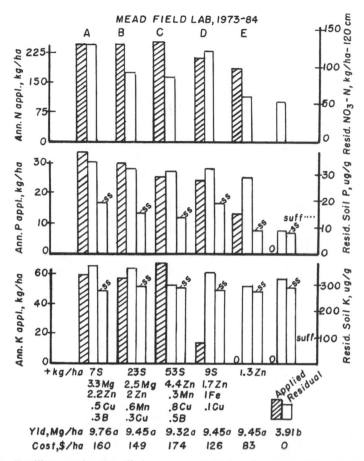

Fig. 5-1. Fertilizer nutrients applied as recommended by five laboratories (A, B, C, D, commercial; and E, Univ. of Nebraska) in relation to yields obtained, fertilizer costs, and nutrient residuals in the soil after the 1984 crop. University of Nebraska-Lincoln Agricultural Field Station, Sharpsburg silty clay loam (Typic Argiudoll), 1973 to 1984 (suff. = sufficiency level; ss = subsoil 15 to 120 cm; yields followed by the same letter are not significantly different at the 5% level).

experiment stations for the first 9 years with extension onto additional sites in the last 4 years. Soil samples were sent to each laboratory each year from plots assigned to that laboratory under a local farmer's name so that the samples would be processed and recommendations made as would be done for a farmer's field. Recommendations were requested for corn (*Zea Mays* L.) with a specified yield goal. The basic objectives were to determine how well the agronomic, economic, and environmental interests of agriculture in the state were being served by soil testing and to further verify the adequacy of university recommendations for satisfying those interests.

Twelve- and 11-year results for the study at the Mead, Northeast, and North Platte stations are presented in Fig. 5-1, 5-2, and 5-3, respectively. Figure 5-4 depicts results for the 4-year study on the Scottsbluff station. Very

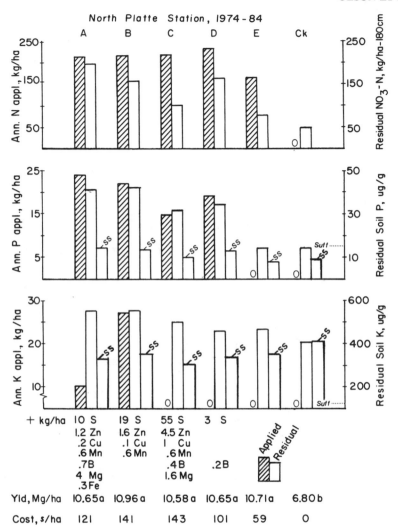

Fig. 5-2. Fertilizer nutrients applied as recommended by five laboratories (A, B, C, D, commercial; and E, Univ. of Nebraska) in relation to yields obtained, fertilizer costs, and nutrient residuals in the soil after the 1984 crop. North Platte Research and Extension Center, Cozad silt loam (Fluventic Haplustoll), 1974 to 1984 (suff. = sufficiency level; ss = subsoil 15 to 120 cm; yields followed by the same letter are not significantly different at the 5% level).

apparent at each of these sites is a wide variance in kinds and amounts of nutrients recommended by the different labs but no difference in yields obtained. Correspondingly, the cost of nutrients applied by the commercial recommendations average somewhat over twice that of the University's. Little question exists from these four long-term comparisons that the sufficiency approach has been adequately serving the farmer's best economic interests.

The same figures depict changes that have occurred in soil test levels of N, P, and K from application of the fertilizers recommended; all ana-

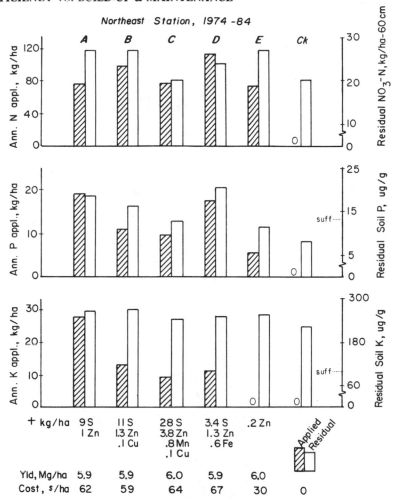

Fig. 5–3. Fertilizer nutrients applied as recommended by five laboratories (A, B, C, D, commercial; and E, Univ. of Nebraska) in relation to yields obtained, fertilizer costs, and nutrient residuals in the soil after the 1984 crop. Northeast Research Extension Center, Moody silt loam (Udic Haplustoll), 1974 to 1984 (suff. = sufficiency level; yields followed by the same letter are not significantly different at the 5% level).

lyzed uniformly at one time in the University lab on samples collected at the beginning of the 1985 season. In the case of N, enough NO_3-N (nitrate-nitrogen) had accumulated in the rooting profile of soil from some of the recommendations to support an optimum yield the following year with limited additional N required.

Soil P levels likewise have grown substantially from some of the laboratories' recommendations, modestly by others. In some cases the soil test value has grown to two to three times that of the "sufficiency level" yet the use of P is still recommended by those laboratories. Certainly the applied P is not being lost in the process, but its continued use at such high

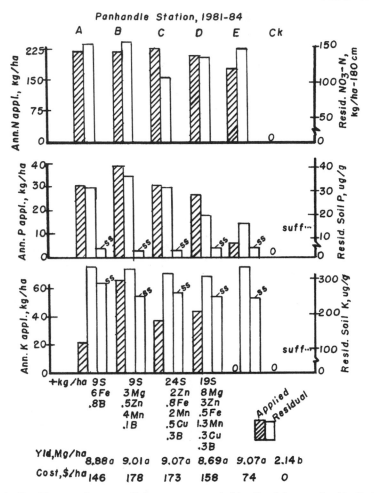

Fig. 5–4. Fertilizer nutrients applied as recommended by five laboratories (A, B, C, D, commercial; and E, Univ. of Nebraska) in relation to yields obtained, fertilizer costs, and nutrient residuals in the soil after the 1984 crop. Panhandle Research and Extension Center, Tripp very fine sandy loam (Aridic Haplustoll), 1981 to 1984 (suff. = sufficiency level; ss = subsoil 15 to 180 cm; yields followed by the same letter are not significantly different at the 5% level).

test levels could contribute to cash flow problems for the farmer. Of interest in Fig. 5–1 and 5–2 is that subsoil P values have increased in plots receiving the higher P rates, probably due to root translocation from the surface. This was not apparent at the Panhandle site (Fig. 5–4), where the subsoil is highly calcareous.

No appreciable change in soil exchangeable K appeared at any of the sites despite the rather liberal K applications advocated by some of the labs, none by others. We can attribute this to the very high levels of inherent exchangeable K of these soils into which the applied increments have disappeared. Fertilizer K rates obviously have had very little impact on the mineral, exchangeable, and water-soluble components of the K equilibria systems existing in these high K level soils. They are likely to remain high in exchangeable K for a long time because most of the K taken up by the crop is returned in its residue to the soil.

These residual effects give evidence that soil N and P levels are building up excessively with the build-up/maintenance approach. It would seem only a matter of time until deleterious effects might be observed from interaction with other nutrients, e.g., P $\times$ Zn or PZn $\times$ Fe (Stukenholtz et al., 1966; Watanabe et al., 1965). Of particular interest is the fact that the sufficiency method is slowly increasing or at least maintaining soil test levels of N, P, and K. Thus, the sufficiency method cannot be accused of being responsible for mining the soil.

The other shorter-term soil test comparison experiments conducted on farmers' fields have shown similar results to those of Fig. 5-1 to 5-4 (Agronomy Soil Fertility Staff, 1984). Because of changes in farm programs and farmers' plans it has rarely been possible to continue these for more than 2 or 3 years. They have served the purpose, however, of demonstrating that the principles cited from the long-term studies on experiment stations apply equally on other soils of adjacent farmland.

OTHER LONG-TERM STUDIES

An 11-year N rate study with corn on Cozad silt loam (Fluventic Haplustoll) at the North Platte Station, immediately adjacent to the soil test comparison experiment, affords data highly relevant to the issue at hand. Figure 5-5 depicts a cost/return relation with maximum profit accompanying little more than 112 kg N/ha applied. If this field had been fertilized by an N replacement of 21.4 N/kg (1.2 lb N/bu) of corn with a yield objective of 10 700 kg/ha (170 bu/acre), the recommended rate would have been 229 kg N/ha. Certainly, it would have achieved the yield objective but with a much lower economic return than from half that rate of N. Furthermore, the slightly higher yield would have been accompanied by a substantial leakage of NO_3-N from the root zone for possible eventual groundwater contamination (Fig. 5-6). This environmental hazard was predicted for Nebraska from the residual NO_3-N measured in soils of numerous field trials conducted prior to 1964 (Olson et al., 1964).

Something analogous to Fig. 5-6 has been taking place in major agricultural production regions of the country in recent times. Because fertilizer N is inexpensive in relation to its impact on yields compared to other inputs, farmers have tended to overuse it for insurance purposes. If not for that reason, perhaps they have been overly optimistic in projecting their yield objectives. Whichever the reason, the continued application of substantially more fertilizer N than is being harvested in crops will continue to contribute to unwanted N accumulation in a region's water resource over time.

A corresponding 10-year study involving rates of P and K for irrigated corn and dryland wheat (*Triticum aestivum* L.) on four major soils of

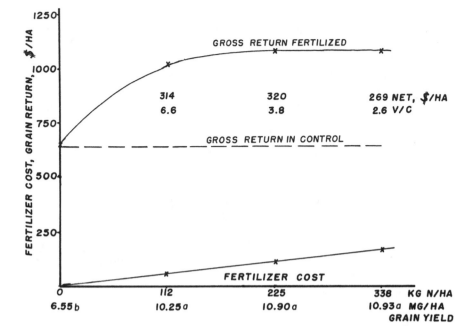

Fig. 5-5. Fertilizer cost and net return in irrigated corn with varied rates of N fertilizer, 1974 to 1984 (corn values at U.S. $0.10/kg and fertilizer N at U.S. $0.50/kg; V/C = value/cost ratio, i.e., $ return/$ of fertilizer; Grabouski, North Platte Research and Extension Center; yields followed by the same letter are not significantly different at the 5% level).

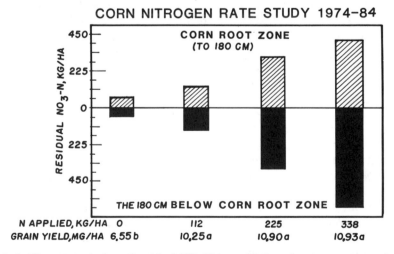

Fig. 5-6. The accumulation of residual NO_3-N in and below the corn root zone in irrigated Cozad silt loam in relation to rate of fertilizer N applied and grain yields obtained (yields followed by the same letter are not significantly different at the 5% level).

Nebraska revealed a small yield increase from applied P at the 22 kg/ha rate (Olson et al., 1982b). Soil test P was being maintained by approximately 11 kg of P applied annually, moderately increased by 22 kg, and almost tripled by 33 kg in all four soils. The maintenance approach in recommendations calling for 2.7 g P/kg (0.15 lb P/bu) of corn at the realized yield of 10 700 kg/ha (170 bu/acre) with one irrigated soil would have called for 29 kg P/ha annually, which would have more than doubled soil test P level in the 10-year period. Comparable results were obtained with the three nonirrigated soils.

Accumulation of soil test K, on the other hand, was barely detectable by the nominal K application rates of 28 and 56 kg K/ha on these high K soils. Rather, there was a slight increase in exchangeable K even without K applied on three of the soils, presumably due to release of fixed K by the substantial amount of NH_4 fertilizer applied, and/or to K extracted from subsoil horizons induced by N stimulation of crop root activity in the subsoil and subsequent deposition in the surface with crop residues. The maintenance concept would have called for around 40 kg K/ha to be applied annually for the projected 10 700-kg irrigated corn yield.

Conceptual differences exist between the sufficiency vs. build-up and maintenance approaches as the former accents fertilizing the crop and the latter emphasizes fertilizing the soil, techniques that are strongly impacted by the timing and placement of fertilizers. With mobile nutrients like N, the time of application that coincides with the period of maximum crop uptake of the element allows a notably lower rate of fertilizer for optimum yield, than in a situation where N is applied earlier (Olson et al., 1964). Likewise, row application or deep band application of immobile nutrients like P and K will commonly require half or less the rate for optimum yield that is necessary with broadcast application (Peterson et al., 1981).

ENVIRONMENTAL IMPLICATIONS

There are many sources for the nutrients that contaminate surface water bodies and groundwater. These include human and animal wastes; atmospheric fallout of inorganic N emitted from plants, animals, and their residues; emissions of NO_x from factories and internal combustion engines; the erosion of fertile soil materials; and the mineralization of nutrients from the native soil organic matter, as well as inorganic fertilizers. Sorting out the major causative factor is not easily done where pollution has occurred, especially with nonpoint sources.

From the authors' viewpoint the build-up and maintenance approach to soil testing and fertilization is certain to accentuate environmental pollution by nutrients compared with the sufficiency approach. The accompanying Fig 5–1 through 5–4 give evidence of excessive nutrient accumulations in soil by build-up/maintenance recommendations, which with run-off, will aggravate eutrophication problems in surface water and with deep percolation, the contamination of groundwater with NO_3. Problem areas of groundwater contamination are recognized countrywide and are receiving increasing public attention (CAST, 1985; Frank et al., 1984; NRC, 1978). Bills have even been under consideration by legislative bodies that would impose control on use of fertilizer nutrients by farmers that, if passed, could have a disastrous impact on the agricultural productivity of the USA.

Tracer N investigations have demonstrated that fertilizer N applied at the most economic rate by the most appropriate method for irrigated corn will leave little excess for groundwater pollution (Broadbent & Carlton, 1978; Olson, 1979). Figure 5–6 gives a similar indication. The fertilizer N rates needed for optimizing yield in the above-cited studies would never accompany build-up and maintenance recommendations. Correspondingly, quite precise calibrations have been developed for P and K soil tests that, when followed, will produce maximum economic yields while still gradually increasing soil test values until no further fertilizer is recommended. This invaluable background research to soil test calibration and interpretation must not be discounted if nutrient contamination of the environment is to be controlled.

FUTURE NEEDS

Continued and expanded soil test calibration work will be needed in the future, especially in conjunction with soil fertility trials with the objective of achieving maximum fertilizer use efficiency. The more completely fertilizer nutrients are utilized by crops the more economic they will be for the farmer and less will be left over as potential pollutants. With changing conditions of crop varieties/hybrids, cropping practices, the environment, etc., yield potentials will vary with attendant changes in calibration for at least some nutrients. Further studies will undoubtedly develop improved methods of extracting soil nutrients for which new or adjusted calibrations will be necessary.

Serious thought and study must be devoted to the issue of soil sampling depth. By so doing, the contribution of subsoil nutrient reserves to crop feeding will be better understood and brought into the refined test calibration scheme. We have substantial evidence of the need for deep sampling of soils for mobile nutrients like N, but the impact of varied levels of soil profile P and K must also be integrated into the ultimate interpretation of test results. This is being accomplished in some degree by those laboratories integrating soil mapping units into the calibration system.

Additional work is needed on the matter of handling and processing soil samples. It seems quite probable that tests for several soil nutrients are best judged by the nutrient response of a crop growing on the soil if field-moist samples are analyzed rather than completely air-dried samples. The latter are much easier to handle and store for periods of time compared to the logistics required for field-moist samples, but the oxidation/reduction status of some elements is radically changed in the air-drying process.

Early work has demonstrated that time of soil sampling is a significant factor affecting test results with some elements. As CO_2 concentrations of the soil-air increase during the growing season from root respiration, soil pH declines with an attendant impact on soil nutrient equilibria. Calibrations need to be refined with incorporation of such variables into the computer algorithms that interpret test values and determine the rate of fertilizer nutrient to be applied.

A most difficult question to be resolved for the future is that of how soil samples are to be procured from fields where minimum tillage and deep row banding of fertilizer were accomplished in prior years, measures increasingly adopted by farmers. Whether or not the sampling device encounters the row of banded fertilizer will result in a very high or much lower test value. Also, the immediate soil surface is changed, notably in its chemical and biological properties by a substantial organic residue build-up. Perhaps a resolution will be to sample the soil only after a periodic thorough mixing of the soil, crop residue, and residual fertilizer by plowing or other primary tillage operation.

In those fields with heterogeneous soil and drainage patterns the advisability of grid sampling needs further elaboration. In many cases it may be to the farmer's economic advantage to spot fertilize deficient areas for a period of time until greater nutrient uniformity exists throughout the field.

Clearly there is much more work to be done to refine the practice of soil testing if it is to be accepted as an exact science. Our contention is that soil testing is a science when it employs the sufficiency concept for interpretation of test results and makes recommendations for fertilizer use on our temperate region soils. The maintenance aspect of the build-up/maintenance concept is a comforting expression like country, motherhood, and apple pie but contributes little to the economic well-being of the farmer and can place the 3% of those directly involved in agriculture at odds with the rest of society for environmental nutrient contamination. As the conditions responsible for the recent CAST report (1985) would imply, it is time for the agricultural community to fully implement those measures required for protecting environmental integrity while ensuring the most economic yields to the farmers.

SUMMARY

Discrepancies in soil test recommendations to farmers result primarily from differing soil testing philosophies. The sufficiency concept establishes cut-off levels above which no fertilizer is recommended, whereas the build-up/maintenance approach calls for a rapid build-up to a high soil test plus annual replacement of an amount the crop is likely to remove regardless of the soil test level. Nebraska trials applied by the University's laboratory and commercial soil testing laboratories over the past 12 years comparing the two systems reveal no significant yield differences but do reveal large differences in kinds, amounts, and costs of fertilizers. The sufficiency approach, although more conservative, is nonetheless increasing soil test levels such that sufficiency will eventually result in the least probability of environmental contamination by nutrients. The build-up/maintenance concept does not take into account the capacity of the majority of temperate region soils to supply significant amounts of most nutrients from native mineral reserves for an indefinite period, having its most likely application in tropical and subtropical regions of highly weathered soils. Further resolution is needed for both interpretation systems; on the time and depth of soil sampling, the logistics of sample collection with reduced tillage and banding of fertilizer, and the question of field moist vs. dry samples.

REFERENCES

Agronomy Soil Fertility Staff. 1984. A comparison of suggested fertilizer programs obtained from several soil test laboratory services. Agron. Dep. Rep. 44. University of Nebraska, Lincoln.

Broadbent, F.E., and A.B. Carlton. 1978. Field trials with isotopically labeled nitrogen fertilizer. p. 1–41. *In* D.R. Nielsen and J.G. MacDonald (ed.) Nitrogen in the environment, Vol. I. Academic Press, New York.

Council for Agricultural Science and Technology (CAST). 1985. Agriculture and groundwater quality. Rep. 103. Council for Agricultural Science and Technology, Ames, IA.

Frank, K.D., T. Bockstadter, C. Bourg, G. Buttermore, D. Eisenhauer, and K.F. Dyer. 1984. Nitrogen and irrigation management, Hall County Water Quality Project, University of Nebraska Coop. Ext. Serv., Lincoln.

National Research Council (NRC). 1978. Nitrates: An environmental assessment. National Research Council, National Academy of Sciences, Washington, DC.

Olson, R.A. 1979. Isotope studies on soil and fertilizer nitrogen. p. 3–32. *In* Isotopes and radiation in research on soil–plant relationships. IAEA-SM-235/48, International Atomic Energy Agency, Vienna, Austria.

Olson, R.A., A.F. Dreier, C. Thompson, K. Frank, and P.H. Grabouski. 1964. Using fertilizer nitrogen effectively on grain crops. Nebraska Agric. Exp. Stn. SB 479.

Olson, R.A., K.D. Frank, P.H. Grabouski, and G.W. Rehm. 1982a. Economic and agronomic impacts of varied philosophies of soil testing. Agron. J. 74:492–499.

Olson, R.A., G.W. Rehm, F.N. Anderson, and E.G. Gatliff. 1982b. Rate of soil P and K build-up/decline with intensive crop production on Nebraska Mollisols. Agron. Abstr. American Society of Agronomy, Madison, WI, p. 217.

Peterson, G.A., D.H. Sander, P.H. Grabouski, and M.L. Hooker. 1981. A new look at row and broadcast phosphate recommendations for winter wheat. Agron. J. 73:13–17.

Stukenholtz, D.D., R.J. Olsen, G. Gogan, and R.A. Olson. 1966. On the mechanism of phosphorus–zinc interaction in corn nutrition. Soil Sci. Soc. Am. Proc. 30:759–763.

Watanabe, F.S., W.L. Lindsay, and S.R. Olsen. 1965. Nutrient balance involving phosphorus, iron and zinc. Soil Sci. Soc. Am. Proc. 29:562–565.

6 Soil Test Interpretations: Basic Cation Saturation Ratios and Sufficiency Levels[1]

Donald J. Eckert[2]

The soil testing-fertilizer recommendation process involves several discrete but somewhat interrelated steps including field sampling, sample preparation, chemical analysis, interpretation of analytical results, and development of recommendations based on these results. Although the philosophies and procedures used to perform each of these steps can affect the nature of the ultimate recommendation, it seems that wide discrepancies in fertilizer recommendations developed by different laboratories operating in a given region are most often due to differing philosophies of interpretation.

Soil test interpretation is the process by which one assesses the fertility status of a field based upon a set of known factors. At minimum, it involves an evaluation of chemical test results in terms of basic soil–plant relationship phenomena. More complex interpretations may include consideration of factors such as crop to be grown, yield goal, crop quality, individual soil characteristics, and climate, while in recent years, tillage system and environmental quality have also received increasing attention.

Currently, two philosophies of interpretation are dominant in the soil testing arena. The "sufficiency level" promotes the concept that there are definable levels of individual nutrients in the soil below which crops will respond to added fertilizers with some probability and above which they likely will not respond. The other dominant concept, the "basic cation saturation ratio" (BCSR) philosophy promotes that maximum yields can be acheived by creating an ideal ratio of calcium (Ca), magnesium (Mg) and potassium (K) in the soil. This latter concept, by definition, does not address nitrogen (N), phosphorus (P) or micronutrients (at least directly). Neither addresses nutrients whose levels in the root zone between the times of soil sampling and crop growth may change considerably (i.e., N in humid region soils). Both, however, have continued to receive attention and widespread use in recent years.

---

[1] Salaries and research support provided by state and federal funds appropriated to the Ohio Agricultural Research and Development Center, The Ohio State University, Wooster, OH 44691. Journal Article no. 40–86.

[2] Associate Professor, Department of Agronomy, Ohio State University, Columbus, OH 43210.

Table 6-1. Number of public universities using selected soil test interpretations.

Nutrient	Sufficiency level	Basic cation saturation ratio	Combination	Other
		no. of universities†		
P	42			4
K	41	2		4
Ca	22	4	5	4
Mg	26	5	8	3
Zn	31			12
Fe	15			6
B	20			8
Cu	20			6
Mn	24			10

† Forty-three universities responding.

McLean (1977) presented evidence that, as of 1977, the BCSR concept was the predominant philosophy used by private soil testing laboratories in the North Central region, whereas most university labs in the same region preferred sufficiency level interpretations. A similar trend involving private labs was noted by Liebhardt (1981). As a part of the preparation of this paper, a survey was sent to extension agronomists nationwide, to assess the current status of soil test interpretation at public universities. Forty-three universities responded, 38 of which operated soil testing labs. Results are shown in Tables 6-1 and 6-2.

The majority of respondents indicated that some form of sufficiency level concept guided their interpretations and recommendations. Cation ratios were used by several states in making recommendations for Ca, Mg, and K, whereas others made no recommendations for Ca and Mg, other than assuming that crop needs would be satisfied by liming. Several states used a combination sufficiency-ratio approach such as the one used in Ohio that suggests a minimum level of soil Mg for sufficiency, modified to maintain soil Mg levels at twice soil K levels. Micronutrient interpretations and recommendations were generally based on sufficiency level concepts (often modified

Table 6-2. Sources of information for developing soil fertility recommendations at public universities.

Nutrient	Research		Literature
	In-state	Neighboring state	
		no. of universities†	
P	43	23	10
K	43	25	10
Ca	22	14	11
Mg	25	20	13
Zn	29	21	16
Fe	16	12	10
B	25	17	14
Cu	10	11	16
Mn	22	13	14

† Forty-three universities responding.

by soil pH); however, several states indicated that micronutrient recommendations were based on plant tissue analysis, not soil tests.

All states had conducted research programs leading to P and K recommendations, though recommendations were often modified or reinforced using data from other states (Table 6-2). For other nutrients, however, there seemed to be as much or more reliance placed on data collected in other states. A number of states had no data and made no recommendations for certain micronutrients, particularly iron (Fe) and copper (Cu).

The survey did not request information regarding N. Many states in low rainfall areas do offer N fertilizer recommendations based on soil tests; however, this practice is not common in higher rainfall areas of the eastern USA due to the instability and relatively unpredictable persistence of plant-available N forms in the soil under such conditions. However, several eastern states did indicate a renewed interest in developing N soil tests and in making recommendations based upon them.

THE BASIC CATION SATURATION RATIO CONCEPT

The basic cation saturation ratio concept originated in New Jersey and was the product of the efforts of Bear and his co-workers (Bear et al., 1945; Bear and Toth, 1948; Hunter, 1949; Hunter et al., 1943; Prince et al., 1947). These investigators proposed an "ideal ratio" for saturation of the cation exchange complex of 65% Ca, 10% Mg, 5% K and 20% hydrogen (H). These saturations were derived from approximately 8 years of work with alfalfa (*Medicago sativa,* L.) on New Jersey soils. The concept was modified somewhat by Graham (1959), who proposed saturation ranges of 65 to 85% Ca, 6 to 12% Mg, and 2 to 5% K for Missouri soils, ranges that allow for rather wide variations in actual Ca/Mg/K ratios in the soil. As such, while suggesting cation saturation as a basis for fertilization, Graham did not seem to promote the existance of one "ideal ratio" per se.

The concept of developing soil test interpretations on the basis of cation saturation seems reasonable in light of basic cation exchange phenomena and the effects that the degree of saturation of one cation may have on the availability of itself and other cations [see McLean (1977) for a review of concepts supporting this philosophy]. Since its inception, however, Bear's proposal has met with more criticism than actual evaluation. Studies evaluating the efficacy of interpreting soil tests on the bases of cation ratios or saturations for agronomic crops are relatively few and although no study of which this author is aware has confirmed the existence of an ideal ratio, an evaluation of the results obtained from several of these studies indicates that Bear's concepts may have some merit, as long as one does not become overly concerned with the ideal ratio itself.

Soil pH is often related to percentage base saturation and is raised to acceptable levels by additions of lime, a Ca- based soil amendment. Lime raises soil pH by neutralizing acidity (H), but maintains the increased pH by increasing the percentage base saturation of Ca (plus Mg if dolomitic lime

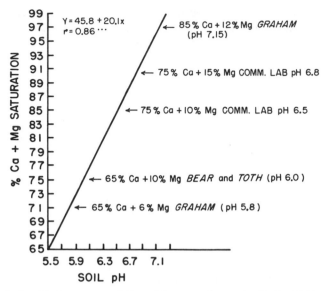

Fig. 6–1. Relationship between total Ca + Mg saturation and soil pH (Liebhardt, 1981). Reproduced from *Soil Science Society of America Journal,* Volume 45, no. 3, May–June, p. 548, by permission of the Soil Science Society of America, Inc.

is used). Liebhardt (1981) noted the relationship between pH and base saturation, and found that Ca + Mg saturations of 75% produced a pH 6.0 level on Delaware Coastal Plain soils (Fig. 6–1), which was desirable from the standpoint of the corn (*Zea mays,* L.)-soybean [*Glycine max,* (L.) Merr.] rotation practiced on these soils. However, as the Ca + Mg saturation was increased above this level, soil pH increased and yields were reduced due to development of manganese (Mn) dificiency in the crop (Liebhardt, 1979). Similar growth responses to pH change were seen in a growth chamber study in Ohio using German millet [*Setaria italica* (L.) Beauv.], although the possibility of Mn deficiency was not evaluated (Eckert & McLean, 1981). Liebhardt reported no decrease in yield below pH 6.0; however, in both field and growth chamber studies using an Ohio silt loam soil, yields for several crops decreased when soil pH was lowered from pH 6.0 to pH 5.0, i.e., decreasing Ca + Mg saturation (Eckert & McLean, 1981; McLean et al., 1983). The relationships between Ca + Mg saturation and soil pH were similar between Ohio and Delaware soils, indicating that the total Ca + Mg saturation proposed by Bear (75% Ca + Mg) might indeed yield a desirable pH for crop growth on many soils of the northeastern USA, and that wide deviations from this saturation may induce yield reductions. Such a deviation could occur in approaching the upper limits (i.e., 97% Ca + Mg) proposed by Graham (1959).

The 2 to 5% K saturation range proposed by Graham (1959) was also evaluated in the Delaware and Ohio studies. In Delaware, Liebhardt (1981) found no effect of varying percentage K saturation within this range on corn or soybean yields. McLean et al. (1983) also found no effect in soybean, but did show a decrease in corn yields as percentage K saturation increased from 2.4 to 4.3%. Alfalfa, however, produced higher yields at this higher K level in 1 of 2 years on the same site.

In Ohio and some other states, soil tests are interpreted by giving consideration to soil cation exchange capacity (CEC) (OCES, 1985). As CEC increases, likewise does the soil K level deemed sufficient, a philosophy that follows the concept of Fisher (1974). The soil test K level is deemed sufficient for corn when it reaches $110 + (2.5 \times CEC)$ mg K/kg (CEC expressed in $cmol_c/kg$). Interestingly, this sufficiency level-based interpretation yields K saturations from 2 to 3.5% K on most Ohio soils. This is within Graham's proposed range, though somewhat below Bear's proposed ideal saturation. As such, it does appear that the K saturation concepts of Bear and Graham do produce soil test K levels recognized as adequate in many instances, though such levels are usually identified as quantities of nutrient rather than percentage saturations.

The place of Mg in the concept is that saturations near 10% Mg should be adequate for crop production, but many recommendations for Mg seem based more on animal nutrition than crop yield considerations (McLean, 1977). The exact soil levels of Mg for maximum crop growth and production are not known with any real certainty, though Eckert and McLean (1981) did show better growth of alfalfa at Mg saturations of 8% than at 4%, supporting previous observations of Prince et al. (1947). The situation with regard to ratios is not so clear. Foy and Barber (1958) showed no effect of varying soil Mg/K ratios on corn yields, and several investigators have also shown little effect of varying soil Ca/Mg ratios on crop yields (Hunter, 1949; McLean and Carbonell, 1972; McLean et al., 1983). Eckert and McLean (1981) were able to create Mg-induced Ca deficiency in German millet and alfalfa with Mg saturations near 15% at pH 5; however, proper liming programs would certainly reduce the likelihood of such an occurrence in actual production situations. Liebhardt (1981) showed no effect on corn yields when soil Mg levels were raised from 11 to 19% at a Ca saturation of 48% Ca (pH 5.7). Thus, it would appear that Mg saturations of 10 to 15% may be satisfactory, but that little emphasis need be placed on fine tuning Ca/Mg or Mg/K ratios as long as sufficient quantities of the individual nutrients are provided.

Perhaps it is unfortunate that Bear chose to express his concepts in terms of an ideal ratio rather than "reasonable saturations." The concept of an ideal ratio has never been verified, but has led to a great deal of controversy in soil testing circles. A review of Bear's own ideas (Bear & Prince, 1945) indicates that the ideal ratio was based as much on economic as agronomic consideration, i.e., saturate the necessary portion of the cation exchange complex as much as possible with the least expensive cation. On mineral soils of the eastern USA, approaching Bear's suggested saturations will generally provide a reasonable soil pH and adequate levels of K and Mg for crop growth. It appears, also, that fairly wide variations in actual ratios are of little consequence, as long as gross imbalances are not created. Thus, an approach addressing individual saturations rather than ratios may lead to reasonable soil test interpretations. Unfortunately, many people using the

BCSR concept have chosen to dwell on ratios, rather than individual saturations, an approach that has led to some rather unrealistic fertilizer recommendation strategies, a topic that is addressed at the conclusion of this paper.

THE SUFFICIENCY LEVEL CONCEPT

The sufficiency level concept promotes the idea that a measurable soil
test level exists below which responses to added fertilizer are probable and
above which they are not. This concept possesses as much intuitive appeal
as the BCSR concept, because it is reasonable to assume that the root zone
can be amended with additions of some nutrients to such an extent that it
is capable of supplying the entire needs of the crop without further additions.

There are certain situations, however, in which this concept is not valid.
The most obvious case deals with mobile nutrients. Agronomists have general-
ly agreed that N is so mobile, and forms so transient, in soils of the eastern
USA that determination of sufficiency levels for N is difficult to impossible.
Another exception involves soils that do not maintain acceptable levels of
K in the root zone due to low CEC, i.e., some sandy soils. Such limitations
are generally recognized when sufficiency level interpretations are developed.

Using the sufficiency level concept requires several preexisting conditions. Analytical procedures must be available that extract a nutrient from
a soil sample in proportion to its level of plant availability under field conditions. The values obtained from soil tests then must be correlated with crop
yields under field conditions to determine at which values responses are likely to occur. Hopefully, such correlations are made for different crops on
different soils. Finally, some measure of the buffering capacities of individual
soils is necessary to allow for reasonable soil buildup recommendations,
because the objective of most people making recommendations based upon
sufficiency levels is to build soils up to or maintain them at or above the
sufficiency level through fertilization.

Different regions of the country have recognized differences in soil
characteristics with regard to soil testing and have often adopted different
extractants to measure the same nutrient. For example, available P is determined in the eastern and central Corn Belt using the Bray and Kurtz P_1 test.
Further west, the same quantity is determined using the Olsen $NaHCO_3$ procedure, whereas in the southeastern US it may be determined using newer
extractants in the Mehlich series. Despite the effort that has gone into developing and selecting them, however, extractants and correlations are by no means
perfect. Skogley and Haby (1981) found no extractant that gave soil test K
values well correlated with crop yields in Montana. Rehm et al. (1981) found
little consistent response to applied K in Nebraska in situations where soil
test levels in surface and subsurface samples indicated the likelihood of
response. Many other examples could be cited. More correlation studies, as
well as continued development of extractants are certainly needed.

Some interest has focused on the development of relatively simple soil
tests that give a simultaneous indication of soil fixation and buffering

capacities as well as existing nutrient levels. McLean (1982) has reviewed tests developed at Ohio State University that use modifications of the standard Bray and Kurtz P_1 and neutral ammonium acetate extractants to accomplish these objectives. These new tests can provide information useful in determining rates of fertilizer to add to *individual* soils to raise or maintain soil test P and K levels, rather than relying on the average buildup responses used at present. Such procedures may prove important in the future if interest in maintaining sufficiency levels continues.

The concept of sufficiency levels as a basis for interpreting soil tests is largely credited to the work of Bray in Illinois (Bray, 1944, 1945), which appeared at the same time as Bear's BCSR concept. Bray related crop yields to various fertility treatments using a mathematical function, which is a modification of the Mitscherlich equation

$$\log (A - y) = \log A - c_1 x_1 - c_2 x_2 \qquad [1]$$

where A = maximum obtained yield, y = any observed yield, x_1 = measured soil nutrient level, x_2 = quantity of added nutrient, and c_1 and c_2 = proportionality constants for x_1 and x_2.

When A, y, c_1, and c_2 are determined using field data, the equation can provide two important pieces of information: (i) the sufficiency level itself; and (ii) the amount of fertilizer required to achieve optimum yields if soil test levels fall below the sufficiency level. This equation cannot be solved for an absolute sufficiency level (i.e., x_1 at A), and to compensate for this most laboratories adopt sufficiency levels at 0.95 to 0.99 A. Despite this drawback, however, this approach to soil test interpretation has had a profound effect on present thinking, yielding concepts such as maximum yield, relating yields to soil tests mathematically, and using soil tests to predict probability of response to added fertilizer. Since his original proposals were made, the concepts of Bray have been adopted, expanded, and modified to fit specific situations, as needs have arisen. Many states use different sufficiency levels for different crops, an idea first promoted by Bray in 1945 but not generally accepted at the time. Some states develop sufficiency levels based on subsoil as well as topsoil fertility. Ohio determines sufficiency levels for K on the basis of soil CEC, an idea promoted by Fisher (1974), who also proposed that crop response functions be analyzed on a polynomical rather than logarithmic basis. Other permutations can be found by examining the fertilizer recommendation guidelines issued by various public and private labs.

One key assumption of the sufficiency level concept is that a sufficiency level for a given nutrient can be defined. This assumption is important, because it identifies the soil test level to which soil buildup recommendations are targeted. Such an assumption cannot be taken for granted, however, because the value of the sufficiency level may be affected as much by data manipulation as by the nature of the data themselves. Melsted and Peck (1977) showed that data analyses using quadratic and logarithmic functions could yield different sufficiency levels. Additionally, an equation (Eq. [1] with field-derived constants) relating corn yield under conventional tillage to Bray and

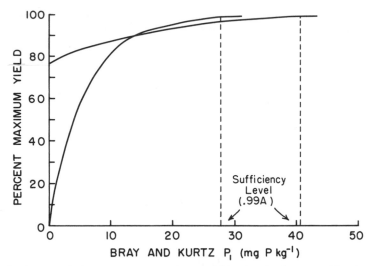

Fig. 6–2. Curves relating percentage maximum yield using the Bray modification of the Mitscherlich equation with and without origin restrictions.

Kurtz P_1 in Ohio (Eckert & Johnson, 1985) generates sufficiency levels of 15 mg P/kg or 24 mg P/kg, depending upon whether a value of 0.95 or 0.99 A is chosen as an operational optimum yield.

There is some reason to consider that this widely used equation may not be totally appropriate for soil test-yield correlation studies as it is written. When expressed graphically to relate percentage yield (as is normally done) to soil test level, this equation generates a curve that passes through the origin (i.e., 0 yield at a 0 soil test level, Fig. 6–2). Acceptance of this phenomenon implies that the extractant used to make the soil test was perfect in extracting the available nutrient from the topsoil sample, that yield can be produced at any measurable soil test level, and that no nutrient contribution is made by the subsoil. None of these assumptions is readily acceptable. Fitting actual data to a logarithmic function in which the data determine the y-intercept produces a different function that gives a different sufficiency level, despite the similarity in curve forms above 0.85 A (Fig. 6–2). The curves in Fig. 6–2 were derived using identical data points from no-tillage P calibration plots (Eckert & Johnson, 1985) and result in sufficiency levels at 0.99 A of 28 and 41 mg P/kg for the origin-restricted and nonrestricted models, respectively. Apparently, the sufficiency level, being an inferred parameter, can be as elusive as the ideal ratio of Bear.

Despite our inability to actually define a sufficiency level itself, the sufficiency level concept possesses one operational advantage that the BCSR concept does not. The large number of studies generated along sufficiency level guidelines has led to the partitioning of soil test results into groups that generally indicate the likelihood or not of a response to added fertilizer. Such a data base does not seem available for the BCSR concept. Fitts (1955) interpreted sufficiency level concepts as a tendency for responses to fertilizer additions to become less probable as soil test levels rise and approach the region of the sufficeincy level. Along these lines, Cate and Nelson (1965) developed graphical methods of partitioning data into low and high probability of response groups, methods that may produce results as satisfying as more complex regression models. Regardless of which method is used, however, the ability to predict probable responses to fertilizer at a given soil test level gives a major advantage to interpretations made using the sufficiency level concept.

IMPACT ON FERTILIZER RECOMMENDATIONS

At present, many fertilizer recommendations made using popular interpretations of both concepts are designed to fertilize the soil rather than the crop itself, and, as such, may not always be in the best economic interest of the farmer. Recommendations derived from the BCSR concept are usually designed to alter the existing soil cation ratio in order to achieve the ideal ratio of 5% K, 10% Mg, and 65% Ca, or to achieve saturations within Graham's ranges regardless of existing or potential yield levels. Likewise, rather than providing only adequate additions of fertilizer to achieve optimum yield at a given soil test level, many recommendations based on sufficiency level concepts today include additional fertilizer to compensate for crop removal and effect a buildup of soil test levels to the sufficiency level, fertilizer that may or may not contribute to profitable yield increases.

Many recommendations using the BCSR concept would be more economically sound and reflective of crop response data if they were restricted to application of K to provide sufficient nutrient, and to applications of lime to maintain sufficient Mg and an acceptable soil pH level. In many cases, such recommendations could probably be made simply on the basis of saturations, rather than ratios. Many people making interpretations, however, may become overly concerned with actual ratios, rather than ensuring that nutrient requirements are being satisfied. Recommendations that are far from cost effective are often made, particularly when one tries to change the Ca/Mg ratio. If the ratio is perceived as being too narrow, recommendations may call for additions of Ca as calcitic lime, even at adequate to high pH levels. Likewise, wide Ca/Mg ratios may generate recommendations for large additions of relatively expensive Mg sources rather than inexpensive dolomitic lime, or recommendations for Mg additions when tissue analysis does not indicate a dificiency. Literature cited previously would indicate that crops tolerate a relatively wide range of Ca/Mg ratios, and there seems to be little justification or benefit to alter an existing ratio unless tissue analysis indicates that nutrient deficiencies do indeed exist. Such alterations can prove to be extremely expensive if they do not produce returns.

Proponents of the sufficiency level concept often make the assumption that optimum yields can be produced only at or near the soil test sufficiency level and that additions of fertilizer at soil test levels below sufficiency will produce increased but not optimum yields (McLean & Watson, 1985; Tisdale

Table 6–3. Rates of P needed to achieve yield goal and Ohio annual P recommendation for corn.

Bray + Kurtz P_1	Yield goal			
	7.5 Mg/ha		9.5 Mg/ha	
	Crop response	Annual recommendation	Crop response	Annual recommendation
mg P/kg	kg P/ha			
5	9	37	48	48
10	0	32	24	40
15	0	22†	1	29†
20	0	22	0	29
25	0	22	0	29
30	0	22	0	29

† Crop removal rate.

& Nelson, 1975). Such statements seem to imply some difference between applied and residual fertility, which is not always borne out by experimental results. Rehm et al. (1981) found maximum experimental yields on low P testing soils with P additions that would not raise soil test levels to sufficiency. Johnson and Wallingford (1983) have also demonstrated that achieving the sufficiency level is not always necessary to produce maximum experimental yields in corn and soybean. In reality, many states seem to accept that the sufficiency level is not always a prerequisite to the most profitable crop production because they offer recommendations that build soil P and K levels to sufficiency over a period of years, rather than immediately.

These "slow buildup" recommendations still often supply more nutrient than is needed to produce a given crop in an "average year." For example, in Ohio, actual rates of P needed to achieve a desired yield goal in most years are often lower than rates recommended by Extension (Table 6–3). The crop response rates in the Table were obtained by solving Eq. [1] with field-derived constants for P for conventional corn (Eckert & Johnson, 1985) using yield goals as y, a procedure that does produce rates which match experimental data quite closely, whereas the standard annual recommendations in the Table (those given to farmers) also include P to effect buildup and compensate for crop removal (OCES, 1985). The higher than crop response rates, which are officially recommended, are often justified on three bases: (i) the popular objectives of raising test levels to the sufficiency level or maintaining them at that level; (ii) the concept of "banking" fertility for future withdrawal; and (iii) building a margin of safety to compensate for differential responses to fertilizers in different years, particularly the need for higher rates in poor-weather years (Johnson & Wallingford, 1983). Such insurance-based concepts are certainly warranted at very low soil test levels, and have generally served farmers well when fertilizer prices were low relative to returns, but changing economic conditions may force some reevaluation of recommendations in the future.

The margin of safety in our and some other's recommendations can be quite wide, and for the farmer experiencing financial difficulties, such insurance may be too expensive. In many cases, rates can certainly be reduced with little danger of reducing yield potential, particularly if existing soil test levels will produce 90 to 100% optimum yield without fertilizer additions. Fertilizer recommendations based on Fitts' (1955) "probability of repsonse" concept may be more appropriate in such situations. Should crop production continue its shift from a long-term to short-term profit perspective, such interpretations will probably become more common.

SUMMARY

Two concepts of soil test interpretation currently dominate the soil testing field: the basic cation saturation ratio (BCSR) and sufficeincy level concepts. The latter is by far the most popular among public laboratories. Current usage of both concepts seems to deviate somewhat from the concepts of their original proponents, a situation that may be leading to over-recommendations for fertilizer in some cases, particularly if crop production is viewed from a short-term rather than a long-term perspective.

Both concepts can provide reasonable fertilizer recommendations, if interpreted properly. However, the BCSR concept follows a rather tortuous path in interpretation and can often generate recommendations that are prohibitively expensive and not justified by agronomic research. Although present interpretations of the sufficiency level concept may also lead to unjustified recommendations in some cases, this concept is backed up by a much larger research base, and is a preferable concept for interpretation of soil test results from both economic and agronomic standpoints. As on-farm soil test levels of available nutrients continue to rise, and crop production becomes oriented more toward short-term profit, "probability of response" interpretations may become dominant concepts in soil testing and fertilizer recommendation.

REFERENCES

Bear, F.E., and A.L. Prince. 1945. Cation equivalent constancy in alfalfa. J. Am. Soc. Agron. 37:217–222.

Bear, F.E., A.L. Prince, and J.L. Malcolm. 1945. The potassium needs of New Jersey soils. New Jersey Agric. Exp. Stn. Bull. 721.

Bear, F.E., and S.J. Toth. 1948. Influence of calcium on availability of other soil cations. Soil Sci. 65:67–74.

Bray, R.H. 1944. Soil-plant relationships: I. The quantitative relation of exchangeable K to crop yields and to crop response to potash additions. Soil Sci. 58:305–324.

Bray, R.H. 1945. Soil-plant relationships: II. Balanced fertilizer use through soil tests for K and P. Soil Sci. 60:463–473.

Cate, R.B., Jr., and L.A. Nelson. 1965. A rapid method for correlation of soil test analyses with plant response data. North Carolina Agric. Exp. Stn. International Soil Testing Series Tech. Bull. no. 1.

Eckert, D.J., and J.W. Johnson. 1985. Phosphorus fertilization in no-tillage corn production. Agron. J. 77:789–792.

Eckert, D.J., and E.O. McLean. 1981. Basic cation saturation ratios as a basis for fertilizing and liming agronomic crops: I. Growth chamber studies. Agron. J. 73:795–799.

Fisher, T.A. 1974. Some considereations for interpretation of soil tests for P and K. Missouri Agric. Exp. Stn. Res. Bull. 1007.

Fitts, J.W. 1955. Using soil tests to predict a probable response to fertilizer application. Better Crops Plant Food 39:17–20.

Foy, C.D., and S.A. Barber. 1958. Magnesium deficiency and corn yield on two acid Indiana soils. Soil Sci. Soc. Am. Proc. 22:145–148.

Graham, E.R. 1959. An explanation of theory and methods of soil testing. Missouri Agric. Exp. Stn. Bull. 734.

Hunter, A.S. 1949. Yield and composition of alfalfa as affected by variations in calcium–magnesium ratio in the soil. Soil Sci. 67:53–62.

Hunter, A.S., S.J. Toth, and F.E. Bear. 1943. Calcium–potassium ratios for alfalfa. Soil Sci. 55:61–72.

Johnson, J.W., and W. Wallingford. 1983. Weather-stress yield loss: Proper fertilization reduces risk. Crops Soils Mag. 35(6):15–18.

Liebhardt, W.C. 1979. Corn yield as affected by lime rate and type on a Coastal Plain soil. Soil Soc. Am. J. 43:985–988.

Liebhardt, W.C. 1981. The basic cation saturation ratio concept and lime and potassium recommendations on Delaware's Coastal Plain soils. Soil Sci. Soc. Am. J. 45:544–549.

McLean, E.O. 1977. Contrasting concepts in soil test interpretation: Sufficiency levels of available nutrients versus basic cation saturation ratios. p. 39–54. *In* T.R. Peck et al. (ed.) Soil testing: Correlating and interpreting the analytical results. Spec. Pub. 29. American Society of Agronomy, Crop Science Society of America, and Soil Science Society of America, Madison, WI.

McLean, E.O. 1982. Use of regression analysis for converting quicktest-measured lime and fertilizer recommendations into actual values based on longer equilibration times. Commun. Soil Sci. Plant Anal. 13:891–897.

McLean, E.O., and M.D. Carbonell. 1972. Calcium, magnesium and potassium ratios in two soils and their effects upon yields and nutrient contents of German millet and alfalfa. Soil Sci. Soc. Am. Proc. 36:927–930.

McLean, E.O., R.C. Hartwig, D.J. Eckert, and G.B. Triplett. 1983. Basic cation saturation ratios as a basis for fertilizing and liming agronomic crops: II. Field studies. Agron. J. 75:635–639.

McLean, E.O., and M.E. Watson. 1985. Soil measurement of plant-available potassium. *In* R.D. Munson (ed.) Potassium in agriculture. American Society of Agronomy, Crop Science Society of America, and Soil Science Society of America, Madison, WI.

Melsted, S.W. and T.R. Peck. 1977. The Mitscherlich-Bray growth function. *In* T.R. Peck et al. (ed.) Soil testing: Correlating and interpreting the analytical results. Spec. Pub. 29. American Society of Agronomy, Crop Science Society of America, and Soil Science Society of America, Madison, WI.

Ohio Cooperative Extension Service (OCES). 1985. Agronomy guide. Ohio Coop. Ext. Service Bull. 472.

Prince, A.L., M. Zimmerman, and F.E. Bear. 1947. The magnesium-supplying power of 20 New Jersey soils. Soil Sci. 63:69–78.

Rehm, G.W., R.C. Sorensen, and R.A. Wiese. 1981. Application of phosphorus, potassium, and zinc to corn grown for grain or silage: Early growth and yield. Soil Sci. Soc. Am. J. 45:523–528.

Skogley, E.O., and V.A. Haby. 1981. Predicting crop responses on high-potassium soils of frigid temperature and ustic moisture regimes. Soil Sci. Soc. Am. J. 45:533–536.

Tisdale, S.L., and W.L. Nelson. 1975. Soil fertility and fertilizers. 3rd ed. Macmillan Publushing Co., New York.

7 Soil Acidity: Soil pH and Lime Needs[1]

M. M. Alley and L. W. Zelazny[2]

Agricultural lime use in 1984, the last year for which figures are available, was estimated at 23.74 million Mg (Remick, 1986). This lime was valued in excess of U.S. $400 million. Estimates have been made that more than three times this amount of lime could be profitably utilized by U.S. farmers (Barber, 1984). Whether or not more lime could be utilized by growers, it is imperative that lime be efficiently utilized in crop production.

Thomas and Hargrove (1984) discuss the current concepts on the chemistry of soil acidity in a concise fashion in the recent Agronomy monograph *Soil Acidity and Liming*; therefore this topic will not be reviewed in this manuscript. In regard to plant response to soil acidity, Foy (1984) indicated that Al^{3+} and Mn^{2+} are the most prevalent toxic ions in acid soils, but in certain cases H^+ may restrict the survival and activity of rhizobia and other soil microorganisms. Moreover, differential plant responses to acid soil conditions are being observed, and selection of plants that tolerate acid conditions are improving crop yields in many regions of the world. This pattern can be expected to continue, particularly with improved plant breeding and genetics techniques. Lime applications will need to be adjusted with the introduction of new cultivars that have improved tolerance to acid soil conditions.

SOIL pH LEVELS FOR EFFICIENT CROP PRODUCTION

The use of lime to neutralize toxic concentrations of Al, Mn, and H^+ and to supply plants with sufficient amounts of Ca and Mg has been practiced for centuries. However, in 1970 a controversy arose concerning the most efficient approach for lime use in agricultural crop production. E.J. Kamprath (1970) proposed that crop yields could be maximized with the addition of lime sufficient to remove Al^{3+} from the exchange sites. This proposal was challenged by E.O. McLean (1970), who stated that lime applications

---

[1]Contribution from the Agronomy Department, Virginia Polytechnic Institute and State University, Blacksburg, VA 24061.

[2]Associate Professor and Professor of Agronomy, respectively.

Table 7-1. Correlation coefficients between relative yield of corn grown in the greenhouse and four measures of soil acidity (Farina et al., 1980).

Correlation (y vs. x)	r	Regression equation†
Percent yield vs. pH_{KCl}	+0.47**	$y = 40.52\ x^{0.51}$
Percent yield vs. pH_{water}	+0.42**	$y = 40.01\ x^{0.49}$
Percent yield vs. percent Al saturation	−0.90***	$y = 96.86 - 0.90x$
Percent yield vs. percent acid saturation	−0.90***	$y = 99.70 - 0.87x$

,* Significant at the 0.01 and 0.001 levels, respectively.
  † y = % maximum yield; x = measure of acidity.

used only to remove exchangeable Al^{3+} ignored other benefits of lime to plants, such as increased N_2 fixation by legumes and enhanced microbial activity. However, after reviewing data from many recent experiments as well as that from earlier work, Thomas and Hargrove (1984) concluded that "recommended pHs should correspond to nontoxic Al^{3+} saturation levels."

Even though the concept of liming to reduce toxicities is accepted in many areas, the idea that a "good soil" has a pH value between 6.0 and 7.0 is still the opinion of many soil scientists, agronomists, and crop producers. Farina et al. (1980) examined exchangeable Al^{3+} and pH as indicators of lime requirements for a range of soils that included two Mollisols, six Ultisols, and one Oxisol. Corn (*Zea mays* L.) was utilized as the test crop, and the data in Table 7-1 illustrate that relative yield was more highly related to Al saturation or acid saturation of the cation exchange complex than either water or salt pH values. Thomas and Hargrove (1984) point out that although such evidence is available, the concept of liming to pH values of 6.0 to 7.0 is perpetuated by the fact that most lime requirement methods must emphasize attaining a certain pH level so that laboratories can process large numbers of samples. This situation is further compounded by growers being advised to keep soil pH levels within certain ranges without being provided information on specific problems.

Given that soil pH is the most commonly utilized indicator of both soil acidity and the need or lack of need for lime, Adams (1984) defined a critical soil pH value as the maximum pH at which liming increases crop yield. He provided an excellent discussion of the development of this concept and stated that "identifying critical pH for a particular crop is needed less for research into causes of acid soil infertility than for advising growers on the need for liming." It is for this very reason that consulting groups and growers will most frequently mention pH values when discussing crop production, whereas researchers will tend to dwell on factors such as Al toxicity, Mo deficiency, and other specific items. Critical soil pH values vary for different crops, and values derived from numerous experiments in the southern USA have been compiled by Adams (1984) and are shown in Table 7-2.

McLean and Brown (1984) in their review of crop response to lime in the midwestern USA stated that "in most cases, crops grown on soils of < pH 5.5 showed some tendency to yield more with additions of lime, but the magnitudes of increase, especially for corn and soybeans [*Glycine max* (L.) Merr.] were not always very large." The previous statement is followed later

Table 7-2. Critical soil pH levels for selected crops grown in the
southern USA (Adams, 1984).

Crop	Critical soil pH
Cotton (*Gossypium hirsutum* L.)	5.0–5.7
Corn (*Zea mays* L.)	5.0–5.5
Soybean [*Glycine max* (L.) Merr]	5.0–5.7
Mo deficient soils	<6.0
Wheat (*Triticum aestivum* L.)	5.5
Peanut (*Arachis hypogaea* L.)	None except as it defines Ca levels
Alfalfa (*Medicago sativa* L.)	6.0
Bermudagrass [*Cynodon dactylon* (L.) Pers.]	5.0
Grain sorghum [*Sorghum bicolor* (L.) Moench]	5.3–5.5

in the review with the comment: "However, the general tendency for yields to increase with pH to ~6.6 is undeniable." The overall idea is that soils in crop production should be limed to at least pH 6.5 to provide a generally more productive environment. This is in agreement with McLean's (1970) concept of liming. Lathwell and Reid (1984) also indicated that soil pH values for optimum crop production should be between pH 6.0 to 7.0 with base saturation values ranging from 60 to 80%. These authors then cite work by Fox (1979) and unpublished experiments by Griffin in Connecticut, which indicated that exchangeable Al saturation was a more accurate indicator than soil pH for predicting reduced corn yields. Hence, we have the paradox of "optimum pH levels" that contrasts with yield responses in liming experiments that are more related to correction of a specific problem, i.e., Al toxicity.

Kamprath's (1984) review of crop response to lime in soils in the tropics points out that the exchangeable cation population in very acid soils of the tropics consists primarily of KCl-exchangeable Al^{3+}, and that the major objective of a liming program on these highly weathered soils is to remove growth limiting factors, not attain certain pH levels. Moreover, as shown by Farina et al. (1980), liming to pH levels near neutrality can result in decreased yields thought to be due to reduced nutrient availability at higher pH levels. Kamprath (1984) summarizes the use of pH in the following statement: "The use of soil pH in diagnosis of acidity problems should be based on the relationship between pH and Al saturation of the ECEC," the ECEC being the effective cation exchange capacity. Aluminum saturation levels of the ECEC that allow for maximum yields on highly weathered Oxisols and Ultisols have been shown to be <10% for wheat (*Triticum aestivum* L.) and soybean crops, whereas corn yields were not restricted with Al saturations <35% (Kamprath, 1984).

LIME REQUIREMENT DETERMINATION

The use of lime to alleviate soil acidity influences on crop yields without creating problems through overliming requires quantification of the acidity present in a specific soil. The most often used methods for measuring amounts of soil acidity are the following (Thomas & Hargrove, 1984):

1. Titration with base or equilibration with lime.
2. Leaching with a buffered solution, followed by titration of the acidity present in the leachate.
3. Subtracting the sum of exchangeable bases from the soil cation exchange capacity (CEC).
4. Equilibration with a buffered solution and estimation of acidity by pH change.
5. Extraction with unbuffered neutral salts and analysis of acidity present in the leachate.

Equilibration of soil samples with a buffered solution and estimation of acidity by pH change is the most widely utilized method to estimate soil acidity. A recent survey by Follett and Follett (1983) revealed that buffer tests for lime requirment are utilized in 28 states. All tests were reported to have strengths and weaknesses, which were summarized by the authors. The most widely utilized tests are the Shoemaker-McLean-Pratt (SMP), Woodruff, Adams-Evans, and Mehlich procedures. All of these procedures were developed with the $CaCO_3$-moisture-incubation method for comparative purposes with no utilization of field calibration experiments (Adams & Evans, 1962; Mehlich, 1976; Shoemaker et al., 1961; Woodruff, 1948). The Woodruff (Brown & Cisco, 1984) buffer method was recently revised but again only incubations were utilized in the revision. The importance and cost of liming for efficient crop production provided the impetus for a Virginia study (Nagle, 1983) to calibrate the most widely utilized lime requirement tests against field pH changes and crop yield responses to lime application. Field lime rate experiments were conducted on seven soils (Table 7–3) in Virginia during the period 1978 to 1982. Lime rates ranging from 0 to 13.44 Mg/ha (Coastal Plain soils) and from 0 to 17.92 Mg/ha (Piedmont and Ridge and Valley soils) were utilized in these experiments. Soil pH change and crop yield response to lime applications were both considered important dependent variables.

The lime requirement tests evaluated in the Virginia study (Nagle, 1983) included the Adams-Evans buffer (Adams & Evans, 1962), the Shoemaker, McLean, and Pratt single buffer (SMP-SB) (Shoemaker et al., 1961) and

Table 7–3. Soil type, classification, physiographic region, water pH, and CEC for soils used in Virginia field studies to calibrate lime requirement tests (Nagle, 1983).

Soil type	Soil taxonomic classification	Physiographic region	Soil-water pH	CEC
				$cmol_c/kg$
Emporia loamy sand	Typic Hapludults	Coastal Plain	4.7	1.7
Goldsboro sandy loam	Aquic Paleudults	Coastal Plain	4.3	2.4
Tomotley loam	Typic Ochraquults	Coastal Plain	4.5	3.9
Masada loam	Typic Hapludults	Piedmont	5.3	3.9
Pacolet sandy clay loam	Typic Hapludults	Piedmont	5.0	2.9
Duffield silt loam	Ultic Hapludalf	Ridge and valley	5.2	3.5
Frederick silt loam	Typic Paleudult	Ridge and valley	5.2	1.9

double buffer (SMP-DB) (McLean et al., 1978), the Yuan double buffer (Yuan, 1974), the Mehlich buffer (Mehlich, 1976), and whole soil titrations. Lime estimates required to reach a soil pH of 6.0 were utilized as a standard basis for the buffer tests because crop yields were maximized at or below that pH level for crops such as corn, wheat, soybean, and alfalfa (*Medicago sativa* L.).

Whole soil titrations are not generally considered to be rapid enough for routine testing programs because 20 to 30 min per sample are required for instruments such as the Radiometer-Copenhagen TT60[3] titration system. Recent advances in instrumentation has reduced this time to 45 to 60 s per sample with a titration system such as the Fisher Titralyzer II.[3] The Virginia study reported a r^2 value of 0.93 when comparing lime requirement predictions to pH 6.0 for rapid vs. slow whole soil titrations of the seven soils described in Table 7-3 (Nagle, 1983). Whole soil titrations are thus a reasonable test to consider if such a measurement provides superior lime requirement predictions.

Regression equations and coefficients of determination for prediction of lime requirement to pH 6.0 vs. observed lime response in the Virginia experiments are shown in Table 7-4 (Nagle, 1983). The whole soil titration to pH 6.0, the SMP double buffer, and the Adams-Evans procedures were equal in terms of predicting lime requirements to pH 6.0. The coefficients of determination for these tests are all much lower than values obtained with greenhouse equilibrations performed during the original development of these tests (Adams & Evans, 1962; McLean et al., 1978; Mehlich, 1976; Shoemaker et al., 1961; Yuan, 1974). Such a reduction in predictive ability would be expected because of greater variation in field situations. The major consideration then is whether or not a coefficient of determination of approximately 0.65 to 0.69 is acceptable for making lime recommendations.

Regression equations and coefficients of determination for the prediction of lime requirements based on observed lime response for 90% max

---

[3] The use of trade names in this publication does not imply endorsement by the authors or Virginia Polytechnic Institute and State University of the products named, or criticism of similar ones not mentioned.

Table 7-4. Regression equations and coefficients of determination (r^2) for prediction of lime requirements based on observed lime requirement to pH 6.0 by six laboratory procedures (Nagle, 1983).

Procedure	Regression equation†	r^2
Whole soil titration (20–30 min per sample) to pH 6.0	$y = 1.9 + 5.4x$	0.65**
SMP-DB	$y = -2.4 + 4.0x$	0.65**
SMP-SB	$y = 0.5 + 1.4x$	0.51**
Adams-Evans buffer	$y = 1.2 + 3.9x$	0.69**
Yuan buffer	$y = -0.1 + 2.6x$	0.38**
Mehlich buffer	$y = -3.0 + 7.3x$	0.60**

** Significant at the 0.01 level of probability.
† y = observed lime requirement (Mg/ha); x = predicted lime requirement with the indicated methods (Mg/ha).

Table 7-5. Regression equations and coefficients of determination (r^2) for prediction of lime requirements based on observed lime response for 90% maximum yield (Nagle, 1983).

Procedure	Regression equation[†]	r^2
Whole soil titration (20–30 min per sample) to pH 6.0	$y = 1.2 + 2.4x$	0.32**
SMP-DB	$y = -0.7 + 1.7x$	0.36**
SMP-SB	$y = 0.9 + 0.5x$	0.18**
Adams-Evans	$y = 1.2 + 1.3x$	0.24**
Yuan buffer	$y = 0.9 + 0.7x$	0.10**
Mehlich buffer	$y = -0.5 + 2.3x$	0.24**

** Significant at the 0.01 level of probability.
[†] y = observed lime requirement (Mg/ha); x = predicted lime requirement with the indicated methods (Mg/ha).

imum yield were also calculated for these six laboratory tests for lime requirement (Table 7-5). The largest coefficient of determination was 0.36, a low value even for field plot experimentation. These low values emphasize the fact that crop response to lime application ends once the particular problem associated with an acid soil condition, i.e., Al toxicity, is corrected. Thus, agricultural advisors must consider the particular problem associated with acid soils of similar parent material and base lime recommendations on eliminating the problem at the lowest cost, instead of assuming that all soils should be adjusted to the same pH level. Specific recommendations should also consider the need for certain pH levels for optimizing other conditions, i.e., herbicide activity, in the crop production system.

SUMMARY

Soil pH is still the most rapid indicator of potential detrimental effects of soil acidity. Soil pH levels for neutralization of detrimental acidity and maximizing crop yields on Oxisols and Ultisols have been shown to be directly related to the removal of exchangeable Al^{3+} and, to a lesser extent, exchangeable Mn^{2+}. The pH level for neutralization of detrimental acidity and maximizing crop yields on Mollisols and Alfisols is believed to be pH 6.5 by many individuals, but crop yield response to lime application data does not support this contention.

Lime requirement buffer tests are well calibrated with greenhouse incubation results but are less well calibrated with field soil pH change. Also, lime requirement buffer tests do not calibrate well with crop response to lime application because yield response to liming is not observed after the specific acidity effect is neutralized. Lime recommendations must be made in the context of solving particular problems rather than merely adjusting all soils to a predetermined pH or "good" level even though laboratory tests must measure acidity to a selected pH level.

Future research should focus on determining the specific problems associated with acid soils in various regions, soil types and management systems, plant species, and the pH level at which the acidity problem is corrected. Lime requirement buffer tests should then be recalibrated for the pH level that eliminates the problem. Such work has been initiated in Quebec with recalibration of the SMP, Yuan, Woodruff, and Mehlich buffers for determining lime requirement for attaining pH 5.5 and 6.0 (Tran & van Lierop, 1982). Finally, the development of rapid titration systems should continue in order to determine the feasibility of utilizing such systems in routine soil testing programs.

REFERENCES

Adams, F. 1984. Crop response to lime in the southern United States. *In* F. Adams (ed.) Soil acidity and liming. 2nd ed. Agronomy 12:211–265.

Adams, F., and C.E. Evans. 1962. A rapid method of measuring lime requirement of red-yellow podzolic soils. Soil Sci. Soc. Am. Proc. 26:355–357.

Barber, S.A. 1984. Liming materials and practices. *In* F. Adams (ed.) Soil acidity and liming. 2nd ed. Agronomy 12:171–209.

Brown, J.R., and J.R. Cisco. 1984. An improved Woodruff buffer for estimation of lime requirements. Soil Sci. Soc. Am. J. 48:587–591.

Farina, M.P.W., M.E. Sumner, C.O. Plank, and W.S. Letzsch. 1980. Exchangeable aluminum and pH as indicators of lime requirement for corn. Soil Sci. Soc. Am. J. 44:1036–1040.

Follett, R.H., and R.F. Follett. 1983. Soil and lime requirement tests for the 50 states and Puerto Rico. J. Agron. Educ. 12:9–17.

Fox, R.H. 1979. Soil pH, aluminum saturation, and corn grain yield. Soil Sci. 127:330–334.

Foy, C.D. 1984. Physiological effects of hydrogen, aluminum, and manganese toxicities in acid soil. *In* F. Adams (ed.) Soil acidity and liming. 2nd ed. Agronomy 12:57–98.

Kamprath, E.J. 1970. Exchangeable aluminum as a criterion for liming leached mineral soils. Soil Sci. Soc. Am. Proc. 34:252–254.

Kamprath, E.J. 1984. Crop response to lime on soils in the tropics. *In* F. Adams (ed.) Soil acidity and liming. 2nd ed. Agronomy 12:349–368.

Lathwell, D.J., and W.S. Reid. 1984. Crop response to lime in the northwestern United States. *In* F. Adams (ed.) Soil acidity and liming. 2nd ed. Agronomy 12:305–332.

McLean, E.O. 1970. Lime requirements of soils—inactive toxic substances or favorable pH range? Soil Sci. Soc. Am. Proc. 34:363–364.

McLean, E.O., and J.R. Brown. 1984. Crop response to lime in the midwestern United States. *In* F. Adams (ed.) Soil acidity and liming. 2nd ed. Agronomy 12:267–303.

McLean, E.O., D.J. Eckert, G.Y. Reddy, and J.F. Triewieler. 1978. An improved SMP soil lime requirement method incorporating double-buffer and quick-test features. Soil Sci. Soc. Am. J. 42:311–316.

Mehlich, A. 1976. New buffer pH method for rapid estimation of exchangeable acidity and lime requirement of soils. Commun. Soil Sci. Plant Anal. 7:637–651.

Nagle, S.M. 1983. Evaluation of selected lime requirement tests for Virginia soils developed through field response of soil pH and crop yields. M.S. thesis. Virginia Polytechnic Institute and State Univ., Blacksburg.

Remick, B.W. 1986. Aglime usage in 1984. National Stone Association, Washington, DC.

Shoemaker, H.E., E.O. McLean, and P.F. Pratt. 1961. Buffer methods of determining lime requirement of soils with appreciable amounts of extractable aluminum. Soil Sci. Soc. Am. proc. 25:274–277.

Thomas, G.W., and W.L. Hargrove. 1984. The chemistry of soil acidity. *In* F. Adams (ed.) Soil acidity and liming. 2nd ed. Agronomy 12:3–56.

Tran, T.S., and W. van Lierop. 1982. Lime requirement determination for attaining pH 5.5 and 6.0 of coarse-textured soils using buffer-pH methods. Soil Sci. Soc. Am. J. 46:1008–1014.

Woodruff, C.M. 1948. Testing soils for lime requirement by means of a buffered solution and the glass electrode. Soil Sci. 66:53–63.

Yuan, T.L. 1974. A double buffer method for the determination of lime requirement in acid soils. Soil Sci. Soc. Am. Proc. 38:437–440.

8   Status of Residual Nitrate-Nitrogen Soil Tests in the United States of America[1]

Gary W. Hergert[2]

Twenty years ago evaluating available nitrogen (N) by the use of mineral N tests was considered to be of limited value in crop production (Bremner, 1965). The importance of mineral N has been proven in many published reports in both semiarid and semihumid regions (Herron et al., 1971; Maples et al., 1977; Olson et al., 1970; Nyborg et al., 1976; Geist et al., 1970). There has been a great deal of research over this time period that has shown that nitrate-nitrogen (NO_3-N) tests can be used very effectively for improving N fertilizer recommendations. These tests are routinely made in drier areas in the western USA and, in many cases, where summer fallow is practiced or in areas where winter or off-season leaching is minimal. They are also heavily used in irrigated areas to estimate carryover NO_3-N that may not have been used by the previous crop. Stewart et al. (1975) prepared a generalized average soil-water percolation map for U.S. corn (*Zea mays* L.) producing soils. The map (Fig. 8-1) shows the potential area for adoption of NO_3-N tests. Most states west of the Mississippi River could use NO_3-N tests if soil test calibration were done. Many of them are currently using NO_3-N tests (Table 8-1, Fig. 8-2). Research on mineral N tests have been thoroughly examined and discussed in recent reviews (Dahnke & Vasey, 1973; Keeney, 1982b). Meisinger (1984) has summarized current soil N evaluation systems used across the USA. The primary soil test used is for NO_3-N (Table 8-1). Some states use NO_3-N and ammonium-nitrogen (NH_4-N). Most of these states have practical guides (extension circulars or fact sheets) explaining sampling procedure and adjustments to their N recommendations when an inorganic N test is used.

One of the primary goals of current soil fertility research is to improve the efficiency of N fertilization. Deep NO_3-N tests have proven their usefulness in the areas where they have already been adopted. The question may be: What areas of the country could improve N recommendations

---

[1] Contribution from Department of Agronomy, Nebraska Agricultural Research Division, Lincoln, NE 68583.

[2] Associate Professor of Agronomy, University of Nebraska West Central Research and Extension Center, North Platte, NE 69101.

Table 8-1. Survey of University/Cooperative Extension soil testing personnel concerning use of NO₃-N Tests.

State†	Test used	Sampling depth§		Status of calibration¶
		Desired	Acceptable for recommendation	
		Western USA		
AZ	NO$_3$-N	0–1.2	0–0.3	Limited work. Use 1960s to 1970s data base.
CA	NO$_3$-N	0–0.9	0–0.6	Ongoing. SB: 7 to 10 year recent data base.
ID	NO$_3$-N	0–1.2	0–0.3	Limited work. Use 1960s to 1970s data base.
NV	NO$_3$-N	0–0.45	?	None. Limited data base.
OR	NO$_3$-N + NH$_4$-N	0–1.8	0–0.6	Ongoing west of Cascades.
		0–0.6	0–0.3	
UT	NO$_3$-N	0–0.9	0–0.3	Limited work and data base.
WA	NO$_3$-N	0–1.8	0–0.6	Ongoing. WW 1980s data.
		0–1.2	0–0.6	Irrigated crops 1970s data base.
		Great Plains USA		
CO	NO$_3$-N	0–0.6	0–0.3	Limited work. SB: 1970s data; WW: 1980s; SG, C: 1950s to 1960s.
KS	NO$_3$-N + NH$_4$-N	0–0.6 or 1.2	0–0.6	Ongoing. 1960s to 1970s data base.
MT	NO$_3$-N	0–1.2	0–0.6	Limited work. 1970s data base.
ND	NO$_3$-N	0–0.6, SB	0–0.6	Ongoing. SB: 1958 to 1980s data base; SG: mid-1970s data base.
NE	NO$_3$-N	0–1.8	0–0.6	Ongoing. WW: 1980s data; C: 1980s data; SB: 1970s data; SB: 1960s data.
NM	NO$_3$-N	0–0.9	0–0.6	Limited work. CT – 1970s data. Use CO & TX data.
OK	NO$_3$-N	0–0.6	0–0.6	Ongoing. WW: 1980s data; Other crops: 1950s to 1960s data.
SD	NO$_3$-N	0–0.6	0–0.6	Ongoing. SG: 1980s data; C: limited data.
TX	NO$_3$-N	0–0.15	0–0.15	Ongoing. 1970s to 1980s data; extending work into more humid areas.
WY	NO$_3$-N	0–1.2	0–0.3	Ongoing. Limited work and data base.

Crops‡ column:
AZ: CT, SG
CA: CT, SB, WW, C
ID: P, SG, SB
NV: B
OR: WW, C / P, Veg.
UT: WW, C
WA: SG / P
CO: SG, P, SB, C, S
KS: W, C, S
MT: WW, SG, B
ND: All NL
NE: WW, SG, SB, C
NM: CT, C, S, WW
OK: All NL
SD: All NL
TX: CT, C, S, WW
WY: All NL

(continued on next page)

Table 8-1. Continued.

State†	Test used	Crops‡	Sampling depth§		Status of calibration¶
			Desired	Acceptable for recommendation	
			Midwestern USA		
AK	NO_3-N	CT	0–0.6	0–0.6	Ongoing. 1980s data base.
IA	--	--	--	--	Beginning. Use NE data for Western IA.
IL	--	--	--	--	Beginning. Research deal more with NO_3 leaching than using residual N.
MI	--	--	--	--	None.
MN (W⅓)	NO_3-N	C, SB, SG	0–1.2	0–0.6	Ongoing for W⅓. Use ND, SD, NE data.
MO	--	CT (boot heel)	--	--	Limited work.
WI	NO_3-N	C	0–0.9	--	Ongoing. C: 4 years of data on corn.

† The following people were contacted in the fall of 1985 to obtain the information presented in this table. AZ, T. Doerge; CA, R. Meyer; ID, R. McDole; NV, D. Thran; OR, J. Hart; UT, C. Topper; WA, A. Halverson; CO, H. Follet; KS, D. Whitney; MT, P. Kresge; ND, E. Vasey; NE, D. Knudsen; NM, L. Cihacek; OK, G. Johnson; SD, J. Gerwing; TX, D. Pennington; WY, K. Belden; AK, C. Snyder; IA, R. Voss; IL, R. Hoeft; MI, M. Vitosh; MN, G. Rehm; MO, D. Bucholz; WI, L. Bundy.

‡ B = barley; C = corn; CT = cotton; NL = nonlegume; P = potato; SB = sugar beet; SG = small grains (oat, spring wheat, winter wheat); S = grain sorghum; WW = winter wheat.

§ Desired means depth (m) requested to provide best recommendation. Acceptable means minimum sample depth required before NO_3-N is used to modify an N recommendation.

¶ Age of data base used for current N recommendations and status of calibration work.

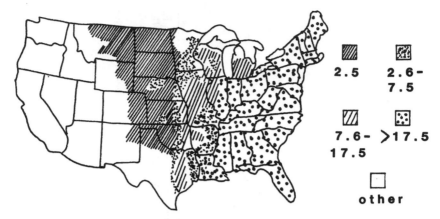

Fig. 8-1. Average annual potential percolation (cm) below the root zone in well-drained soils (Stewart et al., 1975).

through the use of residual NO$_3$-N tests? Research is currently underway in Minnesota, Wisconsin (Malone & Bundy, 1984), and Iowa to expand the use of NO$_3$-N tests (Table 8-1). Much of the research at this point is in the correlation/calibration phase.

Because NO$_3$-N is a mobile and dynamic nutrient, its fate and ultimate utility to the crop is governed by many competing physical and biological factors that interact over time. Site specific variables of soil properties, precipitation, distribution of the precipitation, irrigation management, fertilizer source, and application method can all influence N utilization and in turn the carryover of residual NO$_3$-N. Considering this dynamic nature, it is not surprising that soil testing and N recommendation programs based on residual NO$_3$-N have been difficult to develop and difficult to sell as a sound management technique to farmers and fertilizer dealers. This paper will

Fig. 8-2. Use of residual NO$_3$-N tests by university soil testing laboratories.

discuss three topics related to NO_3-N tests: (i) root-zone sampling including problems of spatial variability, (ii) the challenge of putting together a sufficient data base over a regional or statewide area to develop a successful N recommendation program, and (iii) the impact of implementing such testing procedures over larger areas as it relates to N use efficiency and groundwater or surface water quality. The goal is to increase the overall adoption of soil NO_3-N tests at the farm level.

ROOT-ZONE SAMPLING

The crop root zone is the desired sampling depth for residual NO_3-N; however, the depth necessary to estimate the useable N for the crop, which can be used to modify N fertilizer recommendations, might be less. The depth of sampling required and the accuracy of the field sampling results will be affected by the crop, the soil type, micro- and macroenvironment, and previous history. For practical reasons, the entire root zone may not need to be sampled if sufficient correlation of yield and NO_3-N at shallower depths is high. This will depend on the previously mentioned factors influencing soil NO_3-N content.

CROP CONSIDERATIONS

The rooting pattern of the crop will be the primary determinant in the depth of the NO_3-N sample. For fibrous rooted, full season crops such as winter wheat (*Triticum aestivum* L.), sugarbeet (*Beta vulgaris* L.), and corn grown on well-drained soils, samples to a depth of 1.2 to 1.5 m may be required. For potato plants (*Solanum tuberosum* L.) or other vegetable crops, shallower samples will probably suffice. Local experimentation over several years encompassing a range of local weather patterns is required to determine this minimum sampling depth for a given agroecosystem.

SOIL CONSIDERATIONS

Soil is by nature a heterogeneous body. Biggar (1978) suggested that soils formed on transported materials tend to be more variable than soils derived from bedrock in situ. Jenny (1941) observed that the processes responsible for soil formation are the same processes that govern N distributions. The parent materials are the source of variability but can be drastically modified by climate. The spatial variability of NO_3-N is characterized by an increasing coefficient of variation (CV) as the sampling unit increases. From a research standpoint, it must be remembered that most correlation and calibration to determine the suitability of deep NO_3-N sampling for improving N fertilizer recommendations is done on a small plot basis. When this work is extrapolated to the field for the farmer's use, the size increases dramatically.

Soil properties are not usually normally distributed (Biggar, 1978). A good review of this problem was compiled by Beckett and Webster (1971).

Reuss et al. (1977) determined sampling distribution and variability of NO_3-N in irrigated fields in Colorado and Nebraska. Duplicate cores were obtained from the grid-sampling technique used to determine within grid variability. Reuss et al. (1977) found that the geometric mean represented the NO_3-N content of the sampled fields better than an arithmetic mean. The NO_3-N values changed significantly in parallel and perpendicular directions with respect to the crop rows and were not predictable, indicating that a systematic or grid-sampling plan would be more representative than a random sampling scheme. There seems to be general agreement in the literature that grid sampling is preferable (Peck & Melsted, 1973); however, this method is not usually recommended by most university soil testing services in their Cooperative Extension materials, (Table 8-1) nor what most fertilizer dealers or agricultural consultants are doing. Additional details on sampling are given in chapter 1.

Fertilizer N application generally increases the CV of residual NO_3-N levels compared to unfertilized plots by 100 to 150% (Biggar, 1978). Newer techniques from the discipline of geostatistics concerning the theory of regionalized variables (Matheron, 1971) may provide promise for improving sampling. Application of this technique to agronomic properties has been reviewed recently (Vieira et al., 1983). Russo (1984) discussed designs of optimal sampling networks for estimating the variogram. The proposed sampling network requires an initial sampling network where the points can be located either systematically or randomly although the method's efficiency depends upon the initial sampling network, the sample size, and number of pairs of points per lag.

The study by Russo (1984) does not evaluate the number of sampling points required. A sample size >30 cores per sampling unit is usually recommended (Bresslar & Green, 1982). With a known laboratory variability, other research has found that 36 cores should be composited to obtain a precision of about ±20% of the geometric mean (Reuss et al., 1977). It is estimated that many Colorado farmers composite a maximum of 20 cores per field. In this case, the confidence interval about the geometric mean would be about ±26%. Reuss et al., (1977) concludes that the large number of cores required is disturbing and may represent a limitation in soil sampling programs. If fields test very low in NO_3-N, an error of ±30% is of little consequence as N recommendation will not be affected. In fields with very high NO_3-N levels, an error of this magnitude will also not affect the fertilizer N recommendation. In the mid-range of residual NO_3-N levels [50–150 kg NO_3/(N ha) to 1.8 m], where adjustments are made to N recommendations, the variation could be significant. The work points out the need for further characterization of this problem in the areas where NO_3-N tests are to be used.

Meisinger (1984) used the equation of Harris et al. (1948) to estimate the number of samples required to give an 80% confidence interval (CI) about the mean. The mean CV among a number of fields was 45%. Data showed that compositing 22 cores would normally provide sufficient accuracy to be within the 80% CI. Common soil sampling instructions call for 10 to 20 cores from an area 15 to 20 ha or less. Current sampling procedures were generally developed in an era when fertilizer N use was low and field NO_3-N levels were more uniform. Expanded field sampling research will be required to more accurately and precisely estimate NO_3-N content if the goal is to provide accurate N recommendations to farmers.

ENVIRONMENTAL AND CULTURAL PRACTICE EFFECTS

The climate of a region, the tillage practice, and the irrigation practice will drastically affect the distributions and carryover quantities of NO_3-N. For most dryland crops in the Central Great Plains where residual NO_3-N tests are used, there is evidence that a sampling depth of 0 to 0.6 m is sufficient to evaluate inorganic N status and modify recommendations (Carson, 1980; Ludwick et al.,1977). Ludwick et al. (1977) correlated the NO_3-N present in the 0- to 0.3- and 0- to 0.6-m depths with the amount found in depths of 0 to 1.2 m. The r^2 values were 0.8 and 0.9, respectively. For this reason, Colorado has selected the 0- to 0.6-m depth for NO_3-N sampling. Table 8-1 shows that depending on the region and correlations, a sample from 0 to 0.15 or 0 to 1.8 m may be recommended. As deep soil test for NO_3-N are used in more humid regions of the country, soil type and rainfall distributions will be a greater factor in determining the sampling depth required. Residual NO_3-N does have application some years even in more humid regions (Meisinger et al., 1982).

Recent research in Nebraska (Hergert et al., 1984) has shown that a 0.9- to 1.2-m sampling depth may be appropriate for irrigated or dryland corn. Beyond the initial correlation and calibration of residual NO_3-N tests, sufficient soil sampling studies must be conducted to develop the best sampling plan to ensure an accurate prediction of NO_3-N.

Beyond the scientific aspects of root-zone sampling is the application or practice in the field. Most soil samples today are taken by the private sector, primarily agricultural consultants or local fertilizer dealers. Once sufficient information is developed to recommend a given soil sampling plan, a thorough educational effort will be needed to ensure that the sampling schemes are adopted. More damage to farmer confidence in residual NO_3-N testing may come from a poor soil sample and an underrecommendation of N than could ever be done with the overapplication of 30 or 40 kg N/ha. Sampling equipment that is capable of easily taking a sample from 0.9- to 1.2-m depth is needed by individuals doing the soil sampling. In the long term this may be a more serious limitation in the adoption of deep NO_3-N tests than all of the spatial variability and theoretical considerations previously discussed.

Sample preparation and handling also present problems. Because N transformations can be dramatic if samples are not handled properly, a very thorough effort in educating people taking the samples will be required to guarantee the success of this methodology. Rapid concentration changes from unchecked microbial activity, improper sample handling, or storage [especially the sorption of ammonia (NH_3)] can cause problems. Beyond all of these limitations, however, farmer demonstrations across Nebraska, Colorado, and Kansas (R.A. Wiese, R.H. Follett, & D. Whitney, 1985, personal communication) show that residual NO_3-N tests work and have provided more efficient use of N fertilizer and a more economical fertilizer program. In many situations the use of a good sample with a realistic yield goal will provide a more accurate N recommendation than the use of a yield goal alone. If soil sampling is refined to the point of adequately determining N variability within fields, the next step is to modify fertilizer application equipment to apply variable N rates. With the aid of detailed sampling maps and a microcomputer, a system is already available to do this with dry fertilizer materials (Luellen, 1985). It would be challenge to the equipment manufacturers to develop similar systems for anhydrous NH_3 and N solutions.

GENERATING THE DATA BASE

Recent reports on soil test correlation and calibration have all stressed the importance of developing a sufficiently large data base to improve correlation and calibration (Hanway, 1973; Meisinger, 1984). For calibrating NO_3-N tests, many field sites are required. Each field plot becomes a separate experiment in which the treatment variance and the crop yield are measured by standard methods. However, in the large context of developing an N recommendation algorithm, all of the experimental plots for a given crop are a part of one larger experiment. Similar sets of experimental plots are needed for each crop to develop quantitative relationships. In addition, the experimental locations should be conducted primarily on producers fields in addition to Experiment Station sites. The degree of variability in producers' fields may be somewhat greater than desired; however, this is the area where the fertilizer recommendations will ultimately be used. The data base must be representative of the population to which it will be applied. The survey of soil testing labs (Table 8–1) shows that the data bases currently being used for N recommendations are fairly old and in other cases, nonexistant. At a time when improving N use efficiency and economic return is so important, and when the environmental impacts of so many agricultural chemicals are being questioned, it seems that the agronomic research community has a great challenge to face.

In many instances there is no established funding procedure within the universities or experiment stations for the ongoing conduct of calibration studies to add to or improve an existing data base (J.R. Brown, 1986, personal communication). This may be more of an administrative and/or structural problem than a scientific one, but it does point out that there will be problems in continuing to generate a calibration data base. Calibration research does not appear to be a highly rewarded activity in our existing system because collection of the data base is time consuming and not perceived as a critical need. Administrative techniques will have to be developed to allow for greater staff effort and to generate sufficient funding. It is not a question of whether calibration research is important, it is determining how to marshall the resources to actually get the job done.

As noted by Cope and Rouse (1973), improving the calibration for soil tests should be continuous. In one respect, there is never enough information to make an extremely accurate prediction because only a small sample of the actual existing field population is researched. At some universities there may be a very large data base but it is not sufficiently compiled for review and updating. The use of microcomputers and database management programs make this task easier.

Because the bottom line of crop production is profit, the accuracy of a NO_3-N test and the relationship developed between the test results and crop response depend upon the economics involved. This is especially true now with crop surpluses and the resulting low crop commodity prices farmers are receiving. Since the cost of soil testing for residual NO_3-N is low in relation to the potential returns, a continuous relationship between different levels of NO_3 and the crop response needs to be developed. In most instances, the step from development of the multiple regression equations relating yield to soil test level, to the process of developing a computer algorithm for the actual recommendation is not a well-defined process. The algorithms, recommendation tables, or relationships currently used by most states to make N fertilizer recommendations may bear little resemblance to the regression equations that were developed from the actual research plots.

IMPACT OF WIDE SCALE ADOPTION OF RESIDUAL NO_3-N TESTING

Nitrogen and Environmental Quality

In the long term, improved N use efficiency must be the primary goal of agronomists. Improved N use efficiencies will reduce fertilizer costs and the use of fossil fuel to fix N, and will ultimately influence environmental quality related to N. A number of recent reviews have discussed the impact of N management on minimizing adverse effects on the environment (Aldrich, 1984; Keeney, 1982a; Nielsen & MacDonald, 1978). A recent symposium sponsored by The Fertilizer Institute assessed plant nutrient use and effects on the environment. A number of papers discussed minimizing N losses under different climatic regions across the USA (Broadbendt, 1985; Nelson, 1985; Peterson, 1985; Rhoads, 1985).

PUBLIC PERCEPTION OF FERTILIZER USE

One of the first serious contentions that N may be an environmental threat was discussed by Commoner (1970). Recent studies have confirmed the role of fertilizer N in its relationship to groundwater contamination at

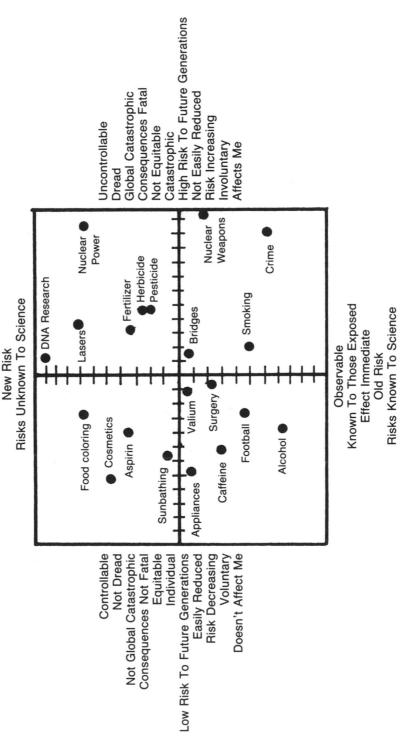

Fig. 8-3. Public perception of risk of dying from current technologies and activities (Allmon, 1985) (Reprinted by permission from the October 1985 issue of *Science 85*. Copyright 1985 by the American Association for the Advancement of Science).

several sites around the USA (CAST, 1985). Some environmental quality problems are related to the use of fertilizer N and often receive publicity. As the general public becomes further removed from the food production process and the percentage of the population actually engaged in farming continues to decrease, a public perception questioning the use of many chemicals used in agriculture has developed (Clark, 1985). In a recent article in *Science 85* a graph was developed that gauged the public's perception of the controllability and manageability of numerous health risks (Allman, 1985). Parts of the graph are reproduced in Fig. 8-3. In the public view, chemical fertilizers are suspect. Most agronomists would agree that fertilizers are very controllable and pose little risk to the public, if handled properly. The public perception, however, makes it imperative that production agriculture make the best use of chemicals and keep them on the targeted area.

IMPROVING NITROGEN FERTILIZER MANAGEMENT

The management of N to limit environmental impact is very simple. Nitrogen that is taken up and fully utilized by a growing crop to provide food, feed, or fiber does not cause unwanted environmental contamination in the current season. The portion of N that remains in the soil in a form that can be lost to groundwater or surface waters does have a potential for degrading water quality. From a simplistic standpoint, applying enough fertilizer N to produce optimum yield of the crop would be the best management strategy. However, the precision of doing that even with all of our research knowledge is not adequate. The yearly variations of climate and soil processes have a drastic influence on N use efficiency. Stewart et al. (1975) demonstrated the large potential for the application of residual NO_3-N tests across the USA. Figure 8-2 indicates wide adoption of soil NO_3-N testing across this area.

Strategies for controlling N input to surface and groundwaters have been outlined recently (Nelson, 1985; Aldrich, 1984). The list as presented by Nelson is given in Table 8-2. To reduce N losses to both surface and groundwater, an appropriate N rate must be applied. Selection of a realistic yield goal combined with root-zone sampling for NO_3-N can greatly improve N

Table 8-2. Best management practices (BMPs) for controlling the entry of N compounds into surface and groundwater (modified after Nelson, 1985).

BMPs for maintaining surface water quality	BMPs for maintaining ground-water quality
1. Apply appropriate N rate	1. Apply appropriate N rate (appropriate yield goal)
2. Timely fertilizer applications	2. Timely fertilizer applications
3. Incorporate fertilizers	3. Improved cropping/irrigation management
4. Proper cropping/residue management	4. Control N transformations
5. Control soil erosion-land structures	5. Foliar applications
	6. Cover crops to scavenge NO_3
	7. Use wastes as N sources

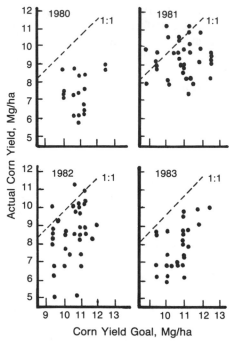

Fig. 8–4. Relationship between yield goal and actual irrigated corn yield for farmer cooperators in the Hall County Nebraska Water Quality Project (Schepers et al., 1986).

use efficiency and reduce potential NO_3-N losses to surface and groundwaters. Across most regions, the fertilizer N rate for grain crops is based on a yield goal that is selected by the producer. Most N recommendations base the N application rate on a factor relating the required amount of N to produce a given amount of grain. If the yield goal is not realistic, the ultimate effect is overapplication of N. Conversely, if the yield goal is too low, the result would be underapplication of N and economic loss.

The difference between adequate or excess N fertilizer is subjective because crop yields vary from year to year due to soil and climatic factors. Prior to planting, however, producers must set a yield goal that will attain a profitable yield regardless of weather conditions. Farmers are generally optimistic about their ability to produce a high yield. A 4-year study in south central Nebraska indicated that only 10% of the producers of irrigated corn reached their yield goal (Schepers et al., 1986). There was year-to-year variation that affected N use efficiency (Fig. 8–4), but approximately 50% of the producers attained only 80% of their yield goal (Fig. 8–5; Schepers & Martin, 1986). On the average, these farmers overestimated their yield goal by 2400 kg/ha. Failure to achieve the yield goal resulted in N applications that averaged 44 kg/ha more than would have been recommended by the University of Nebraska Soil Testing Laboratory if a more realistic yield goal has been used. This research shows the importance of selecting a proper yield goal. As more research is conducted with a specific crop in a specific region, soil

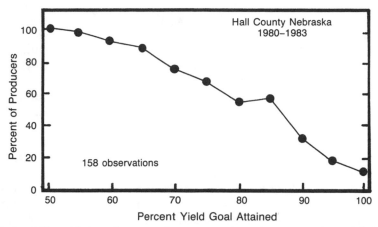

Fig. 8-5. Ability of irrigated corn producers to attain various proportions of their yield goal in the Hall County Nebraska Water Quality Project (Schepers & Martin, 1986).

test calibration may show that yield goal is not the best indicator for making N recommendations. Production capability by soil series or climatological zones may be a better factor for determining N rate and profitable production.

A second way to improve N use efficiency and reduce NO_3-N pollution potential is to use root-zone sampling to determine residual NO_3-N to modify N recommendations. Wide-scale producer use of NO_3-N tests can greatly reduce NO_3-N carryover and thereby reduce the potential pollution and the actual loss of N that may find its way into streams or groundwater. Farmer adoption of this management practice, however, will depend primarily on economic considerations. In many instances root-zone sampling for NO_3-N may mean lower fertilizer N application rates. Fertilizer dealers, commercial soil testing labs, and agricultural consultants must also be convinced because they take more than 80% of the soil samples nationwide. In today's depressed economy there is a golden opportunity for wide-scale adoption of this technique. It should improve environmental quality in the long term, and it may create an immediate economic benefit because of reduced production costs to the farmer.

In most areas where NO_3-N sampling is used there is a large irrigated acreage. Improved irrigation management is still required in many areas to reduce NO_3 leaching (Keeney, 1982a).

Legislation often follows problems of high public concern. Several states are implementing legislation related to increased use of chemicals applied in irrigation water (chemigation). Farmers and agribusiness people would prefer to see increased education as opposed to legislation to control on-farm N use. If sufficient progress is made in the adoption of techniques to improve N use efficiency, legislation may not be necessary. The current economic incentives may be enough to improve the adoption of realistic yield goals and root-zone sampling for NO--N. In many states there are no provisions for encouraging farmers to implement these practices. In times of high grain prices in relationship to cost inputs there is little economic incentive for improving N use efficiency at the farm level.

SUMMARY

The goal of N fertilization should be to obtain near maximum yields, maximize N use efficiency, and to minimize residual soil NO_3 carryover at the end of the growing season (Bock, 1984). There will always be a certain portion of N from the cropping system lost to surface and groundwaters. This occurs because crop production systems are not 100% efficient in removing NO_3-N from soil solution.

Root-zone sampling for residual NO_3-N provides an excellent opportunity for improving N use efficiency. Numerous studies done in an area west of the Mississippi River show that this practice works. Other studies show that farmers generally overestimate their yield goals, which are the basis for most N recommendations. Educational programs to help farmers establish more realistic yield goals will also contribute to improved N use efficiency. The task for the next 10 years will be to convince all commercial concerns to adopt residual NO_3 testing as a standard part of developing N recommendations.

REFERENCES

Aldrich, S.A. 1984. Nitrogen management to minimize adverse effects on the environment. p. 663–676. *In* R.D. Hauck (ed.) Nitrogen in crop production. American Society of Agronomy, Crop Science Society of America, and Soil Science Society of America, Madison, WI.

Allman, W.F. 1985. Staying alive in the 20th century. Science 85. October, p. 31–41.

Beckett, P.H.T., and R. Webster. 1971. Soil variability: A review. Soils Fert. 34:1–15.

Biggar, J.W. 1978. Spatial variability of nitrogen in soils. p. 201–211. *In* D.R. Nielsen and J.G. MacDonald (ed.) Nitrogen in the environment. Vol. I. Academic Press, New York.

Bock, B.R. 1984. Efficient use of nitrogen in cropping systems. p. 273–294. *In* R.D. Hauck (ed.) Nitrogen in crop production. American Society of Agronomy, Crop Science Society of America, and Soil Science Society of America, Madison, WI.

Bremner, J.M. 1965. Inorganic forms of nitrogen. *In* C.A. Black et al. (ed.) Methods of soil analysis, Part 2. Agronomy 9:1179–1237.

Bressler, E., and R.E. Green. 1982. Soil parameters and sampling scheme for characterizing soil hydraulic properties of a watershed. Tech. Rep. 148. University of Hawaii, Honolulu.

Broadbent, F.E. 1985. Minimizing nitrogen losses in western valleys. p. 153–172. *In* Plant Nutrient Use and the Environment, Fertilizer Inst. Symp. Proc., Kansas City, MO. 21–23 October. The Fertilizer Institute, Washington, DC.

Carson, P.L. 1980. Recommended Nitrate-nitrogen tests. p. 12–13. *In* W.C. Dahnke (ed.) Recommended chemical soil test procedures for the north central region. North Cent. Reg. Pub. 221. revised ed. North Dakota Agric. Exp. Stn.

Clark, E.H. 1985. Plant nutrients and environmental impacts. p. 25–47. *In* Plant Nutrient Use and the Environment, Fertilizer Inst. Symp. Proc., Kansas City, MO. 21–23 October. The Fertilizer Institute, Washington, DC.

Commoner, B. 1970. Threats to the integrity of the nitrogen cycle: Nitrogen compounds in soil, water, atmosphere, and precipitation. p. 70–95. *In* S.F. Singer (ed.) Effects of environmental pollution. Springer-Verlag, New York.

Cope, J.T., Jr., and R.D. Rouse. 1973. Interpretation of soil test results. p. 35–54. *In* L.M. Walsh and J.D. Beaton (ed.) Soil testing and plant analysis. Soil Science Society of America, Madison, WI.

Council for Agricultural Science and Technology (CAST). 1985. Agricultural and ground-water quality. Report no. 103. Council for Agricultural Science and Technology, Ames, IA.

Dahnke, W.C., and E.H. Vasey. 1973. Testing soils for nitrogen. p. 97-114. *In* L.M. Walsh and J.D. Beaton (ed.). Soil testing plant analysis. Soil Science Society of America, Madison, WI.

Geist, J.M., J.O. Reuss, and D.D. Johnson. 1970. Prediction of nitrogen fertilizer requirements of field crops: II. Application of theoretical models to malting barley. Agron. J. 62:385-389.

Hanway, J.J. 1973. Experimental methods for correlating and calibrating soil tests. p. 55-66. *In* L.M. Walsh and J.D. Beaton (ed.) Soil testing and plant analysis. Soil Science Society of America, Madison, WI.

Harris, M., D.G. Horvits, and A.M. Mood. 1948. On the determination of sample sizes in designing experiments. J. Am. Stat. Assoc. 43:391-402.

Hergert, G.W., E.J. Penas, G.W. Rehm, and R.A. Wiese. 1984. Improving nitrogen fertilizer recommendations for corn in Nebraska. Agron. Abstr. American Society of Agronomy, Madison, WI, p. 207.

Herron, G.M., A.F. Dreier, A.D. Flowerday, W.L. Colville, and R.A. Olson. 1971. Residual mineral N accumulation in soil and its utilization by irrigated corn (*Zea mays* L.). Agron J. 63:322-327.

Jenny, H. 1941. Factors of soil formation. McGraw-Hill, New York.

Keeney, D.R. 1982a. Nitrogen management for maximum efficeincy and minimum pollution. *In* F.J. Stephenson (ed.) Nitrogen in agricultural soils. Agronomy 22:605-649.

Keeney, D.R. 1982b. Nitrogen-availability indices. *In* A.L. Page et al. (ed.) Methods of soil analysis, Part 2. 2nd ed. Agronomy 9:711-733.

Ludwick, A.E., P.N. Soltanpour, and J.O. Reuss. 1977. Nitrate distribution and variability in irrigated fields of northern Colorado. Agron. J. 69:710-713.

Luellen, W.R. 1985. Fine-tuned fertility: Tomorrow's technology here today. Crops Soils 38(2):18-22.

Malone, E.S., and L.G. Bundy. 1984. Effects of residual inorganic N in soil profiles on corn N fertilizer requirements. Agron. Abstr. American Society of Agronomy, Madison, WI, p. 213.

Maples, R., J.G. Keogh, and W.E. Sabbe. 1977. Nitrate monitoring for cotton production in Loring-Calloway silt loam. Arkansas Agric. Exp. Stn. Bull. no. 825.

Matheron, G. 1971. The theory of regionalized variables and its application. Les Cahiers du Centre de Morphologie Mathmatique de Fontainebleau no. 5. Centre de Geostatistique, Fontainebleau, France.

Meisinger, J.J. 1984. Evaluating plant-available nitrogen in soil-crop system. p. 391-417. *In* R.D. Hauck (ed.) Nitrogen in crop production. American Society of Agronomy, Crop Science Society of America, and Soil Science Society of America, Madison, WI.

Meisninger, J.J., T.H. Carski, M.A. Nys, and V.A. Bandel. 1982. Residual nitrate nitrogen in Maryland. Agron. Abst. American Society of Agronomy, Madison, WI, p. 215.

Nelson, D.R. 1985. Minimizing nitrogen losses in non-irrigated eastern areas. p. 173-210. *In* Plant Nutrient Use and the Environment, Fertilizer Inst. Symp. Proc., Kansas City, MO. 21-23 October. The Fertilizer Institute, Washington DC.

Nielsen, D.R., and J.G. MacDonald. 1978. Nitrogen in the environment. Vol. I and II. Academic Press, New York.

Nyborg, M., J. Newfeld, and R.A. Bertrand. 1976. Measuring crop available nitrogen. p. 102-117. *In* Proc. Western Canada Nitrogen Symp., Calgary, Alberta, Canada. 20-21 January. Alberta Agriculture, Edmonton, Alberta, Canada.

Olson, R.J., R.H. Hensler, O.J. Attoe, S.A. Witzel, and L.A. Peterson. 1970. Fertilizer nitrogen and crop rotation in relation to movement of nitrate nitrogen through soil profiles. Soil Sci. Soc. Am. Proc. 34:448-452.

Peck, T.R., and S.W. Melsted. 1973. Field sampling for soil testing. p. 67-76. *In* L.M. Walsh and J.D. Beaton (ed.) Soil testing and plant analysis. Soil Science Society of America, Madison, WI.

Peterson, G. 1985. Minimizing nitrogen losses in the Great Plains. p.225-246. *In* Plant Nutrient Use and the Environment, Fertilizer Inst. Symp. Proc., Kansas City, Mo. 21-23 October. The Fertilizer Institute, Washington DC.

Reuss, J.O., P.N. Soltanpour, and A.E. Ludwick. 1977. Sampling distribution of nitrates in irrigated fields. Agron. J. 69:588-592.

Rhoads, F. 1985. Minimizing nitrogen losses in irrigated eastern areas. p. 211-224. *In* Plant Nutrient Use and the Environment, Fertilizer Inst. Symp. Proc., Kansas City, MO. 21-23 October. The Fertilizer Institute, Washington, DC.

Russo, D. 1984. Design of an optimal sampling network for estimating the variogram. Soil Sci. Soc. Am. J. 48:708-716.

Schepers, J.S., K.D. Frank, and C. Bourg. 1986. Effect of yield goal and residual soil nitrogen considerations on nitrogen fertilizer recommendations for irrigated maize in Nebraska. Fert. Issue 3:133-139.

Schepers, J.S., and D. Martin. 1986. Public perception of groundwater quality and the producers dilemma. p. 399-411. *In* Agricultural Impacts on Groundwater Conf. Omaha, NE. 11-13 August. Well Water Journal Publishing Co., Dublin, OH.

Stewart, B.A., and D.A. Woolhiser, W.H. Wischmeier, J.H. Caro, and M.H. Freere. 1975. Control of water pollution from cropland. Vol. I. USDA Rep. ARS-H-5-1. U.S. Government Printing Office, Washington, DC.

Vieira, S.R., J.L. Hatfield, D.R. Nielsen, and J.W. Biggar. 1983. Geostatistical theory and application to variability of some agronomical properties. Hilgardia 51(3).

9   Sulfate: Sampling, Testing, and Calibration

Gordon V. Johnson[1]

In many parts of the USA and similar agricultural lands of the world, the need for supplementing natural sources of sulfur (S) to meet plant requirements is unnecessary. Consequently, there is generally little need for S soil testing as a guide for making fertilizer recommendations for crop production. It is important to understand this because the attitude toward sampling, testing, and calibration is strongly influenced by the frequency of S deficiencies. Where S deficiencies do occur in crops, proper sampling and testing, together with the use of appropriate calibration tables, lead to the prudent use of S fertilizers. This paper will present the rationale to support these statements.

DISTRIBUTION OF S IN THE ENVIRONMENT

The importance of sampling, testing, and calibration of S soil tests is closely linked to the behavior of S in soils, water and the atmosphere. In soils both the inorganic (sulfate) SO_4-S and reduced S present in organic combination are important sources of S for plants. Utilization of the latter source is preceeded by mineralization to SO_4. Like nitrate-nitrogen (NO_3-N), SO_4 is a moderately stable chemical form, although there are several different processes, including transformations, that may act upon it. It may remain as SO_4 either as an ion in solution or, unlike NO_3, in combination with calcium (Ca) (or some other predominant cation) as a sparingly soluble salt. As the SO_4^{2-} ion, S is relatively mobile in soils and may be leached out of the surface of permeable soils in areas of high rainfall. Accumulation of SO_4 in the less permeable subsoils often will be an important source of S in these areas. Soil organic matter is an important reservoir of S in all soils. Low levels of available soil S are most apt to occur in very permeable (usually sandy) soils low in organic matter.

The susceptibility of crops to S deficiency in soils low in available S is tempered in many cases by the addition of S from nonfertilizer S sources.

[1] Contribution from Oklahoma State University, Stillwater, OK 74078.
[2] Professor, Department of Agronomy, Oklahoma State University, Stillwater, OK 74078.

Fig. 9–1. Deposition of S (kg S/ha) by rainfall in 1981. (CAST, 1984).

In this regard, a major source of S is the annual deposition by rainfall. Figure 9–1 indicates reasonable expected amounts of S that may be added to soils each year by rainfall. The amount of S deposited by rainfall is influenced by several factors including time, total rainfall, industrial activity, and volcanoes. The smallest amounts are generally associated with low annual rainfall. Soils in these areas usually have inherent high available S levels because the S was not leached out of the earth's mantle during soil genesis. A more recent study (Sharpley et al., 1985) in the Southern Plains of the USA has reported rainfall S deposition ranging from 12 to 30 kg/(ha yr). As will be shown later, deposits of this magnitude may meet 60 to 100% of the S requirement even for high crop yields.

Crop deficiencies of S in desert and arid regions are extremely rare. Soils of these areas developed under conditions where S was not leached out of the rooting depth of most crops. Consequently, the soils are often high in mineral S reserves. The yield potential of these soils without irrigation is low and, hence, the demand from this large resevoir is also small. Most irrigation waters contain significant quantities of SO_4-S, so when the yield potential is raised by irrigation, the higher S requirement is more than met by that in the irrigation water.

TESTING

In order to obtain current information on S soil testing activity, a questionaire was mailed to extension specialists with soil testing responsibilities for each of the 50 states and Puerto Rico.

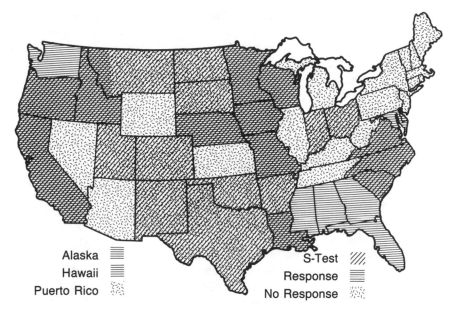

Fig. 9-2. Geographic distribution of S response and/or recommendation of S fertilizers for crop production and the use of S soil tests by land grant institutions.

Twenty-three respondents indicated they offered a soil test for available S and 28 did not. Of the 23 that offer a test, nine do not make recommendations for and/or have not found a response to fertilizer S. The general attitude of this group is probably expressed by one who commented: "We test for S mainly to show that it is not needed very often in spite of the fact that everyone is trying to sell S."

Of the 28 respondents who do not offer a soil test for S, eight indicated they do make a "blanket" recommendation for S, or that response to S is expected under recognized specific crop and soil conditions.

Figure 9-2 shows the geographic distribution of these questionaire responses. Note that except for Kansas, the Great Plains states all offer a S test, although none of them recommend or obtain an S response, except for Nebraska. States along the western Mississippi Valley, except for Arkansas, test for S and make S fertilizer recommendations. Most of the states in the industrialized upper northeast do not offer a S test and do not see S response. Exceptions are the states Indiana and Ohio, which do offer an S test. This group of states receives the highest amount of S in rainfall (Fig. 9-1). The Coastal Plains states of the southeastern USA all observe S response or make general S recommendations. Only Virginia and the Carolinas offer a S test. States not offering an S test usually make S recommendations based on crop and soil type. The response to S is associated with high yielding crops grown in deep sandy soils. Figure 9-2 illustrates general information regarding S testing and responses. However, it should be kept in mind that even though a state is identified as having S responses or making an S recommendation, this does not mean all soils and all crops are tested and respond. For most states the responses are restricted to crops grown on deep sandy soils that do not have a B horizon or a layer of soil with some clay content within 50 cm of the surface.

States in the Rocky Mountain region and western deserts do not make recommendations for S, presumably because the soils have adequate supplies or it is provided in irrigation water. Probably for this reason, Arizona, Wyoming, and Nevada do not test for S. States of the Pacific Northwest generally recommend S for high yielding crops grown in deep permeable soils. Washington does so without the aid of soil testing.

SAMPLING

Seventeen of the states that offer an S soil test rely only on the surface soil sample to provide a measure of available S. Five states include analysis of the subsoil, and one state, South Carolina, tests only the subsoil as a guide to S fertilizer needs. Testing only the subsoil is a novel idea that, given the nature of S in soils, should improve the probability of identifying responsive soils.

Because SO_4-S is relatively mobile in soils, the most accurate estimate of immediately available S must include an analysis of both surface and subsoil. Although there remains some argument for sampling the entire rooting depth for analysis of mobile nutrients, most investigators have given way to reason and practicality and have limited the depth at 60 cm. To a large degree, the S that is not sampled below 60 cm is compensated for by $< 100\%$ utilization of what is found in the upper 60 cm of soil.

Except for the above considerations, testing soils for available S does not present any unique sampling problems or requirements.

EXTRACTION AND ANALYSIS

The most commonly used extracting reagent is monocalcium phosphate at the concentration of 500 mg/kg P. Nine states use this chemical in water, whereas seven use it in 2 M HOAc. Four states used HOAc–NH_4OAc, two use just water, and two use KCl. Other extractants used are NH_4Cl, KCl + KH_2PO_4, and LiCl.

Sulfur in the extract is usually determined turbidometrically with Ba (16 states). Other analytical methods used are: the colorimetric procedure of Johnson and Nishita (1952), ion chromatography, and induction coupled plasma emission spectrometry.

A recent review of procedures by Eik (1980) provides an excellent discussion and comparison of extracting and analytical procedures.

SOIL TEST CALIBRATION

Attempts to identify a soil test value (critical level), above which S response in the field would not be expected, have been met with limited success. Likewise, relating soil test levels from zero to the critical level and different S application rates has seldom been done.

Sulfur Requirement of Crops

The amount of S required by plants is directly related to the amount of protein present and the degree to which it is composed of S-bearing amino acids. The most common indices of plant S adequacy have been SO_4 concentration, total S concentration, and the total N/total S ratio (Freney et al., 1977; Randall & Spencer, 1980; Spencer, 1978; Stewart & Porter, 1969; Westerman et al., 1983). These indices vary considerably depending on crop type, portion of crop sampled, age of plants, and the N and S nutritional status of the plant. The critical level of total plant S has been reported at 0.16% for young wheat (*Triticum aestivum* L.) forage in a greenhouse pot study (Hue et al., 1984) and 0.22% for the first cutting of alfalfa (*Medicago sativa* L.) (Pumphrey & Moore, 1965). Values of 0.17% have been reported for corn (*Zea mays* L.) leaf and 0.14% for coastal bermudagrass [*Cynodon dactylon* (L.) Pers.] (Reneau, 1983; Martin & Matocha, 1973). The ratio of total N/total S, in many of these same studies, will vary from 22:1 to 11:1 as critical values for diagnosing S deficiency.

Although the ratio of total N/total S is most commonly used as an index of plant S status, protein N/protein S has been shown to be more stable (Dijkshoorn & Lampe, 1960; Freney et al., 1977). When total N/total S varied from 4:1 to 331:1, protein N/portein S was relatively constant at about 19:1 in subterranean clover (*Trifolium subterraneum* L.). Removal of nonprotein N and S in addition to special oxidation precedures are apparently necessary to gain constancy in N/S ratios. In an early study of this same concept (Dijkshoorn & Lampe, 1960), it was concluded that the minimal S requirements for normal growth could be calculated by 0.027 × organic N (N/S = 37:1).

It is obvious the S requirement will be underestimated from total plant S measurements if S nutrition has been lacking, and it may be overestimated if available S has been excessive. The N/S ratio has been shown to be a useful indicator of S nutrition, expecially if N nutrition is neither deficient nor excessive. This N nutritional status exists for legumes. Although there remains considerable discrepancy among values reported, a value of 20:1 has support as a critical value when the Cate-Nelson technique (Cate & Nelson, 1965) is used to separate responsive and nonresponsive soils in a recent field study with alfalfa (Nuttal, 1985). Numerous studies in Oklahoma (Bremer, 1975) indicate this may still overestimate the S requirement of forage sorghum [*Sorghum bicolor* (L.) Moench.] and bermudagrass. It seems appropriate then to use a N/S ratio that coincides with adequate (but not excessive) N and S nutrition as a basis for calculating S requirement of all crops.

Table 9-1. Sulfur requirement for wheat and corn grain.

Yield goal		Sulfur
Wheat	Corn	
Mg/ha		kg/ha
1.2	2.2	2
2.4	4.0	4
3.6	5.6	6
4.5	7.1	8
5.3	8.5	10
6.2	9.8	12
	11	14
	12	16
	13	18

Using this approach, Oklahoma has developed tables of S requirements for common agricultural crops (Johnson, 1985). The values shown in Table 9-1 are an example for small grains and were generated by simply dividing the calibrated N requirement (Johnson & Tucker, 1982) for these crops by a factor of 20. One value to this approach is that reliable levels of required S can be identified without extensive field trials. This is especially important for regions where S deficiency is rare, but there is strong concern about adding S. An additional advantage of this approach is that it utilizes local N response data. Many agricultural regions have results of numerous N studies over time on different soils and crops. Hence, their N calibrations are very dependable relative to any S calibration data that might be independently obtained. Two drawbacks to this approach are that: (i) any error in the N calibration (such as making a liberal estimate of N required) is an inherent error in the S calibration; and (ii) it assumes the same crop utilization efficeincy for fertilizer S as for N. Either of these shortcomings are usually vastly outweighed by the value of having reasonable S calibration tables as opposed to none at all.

Interpretation and Recommendations

The SO_4-S soil test is interpreted in much the same way as the NO_3-N soil test. The first step is to identify a realistic yield goal for the crop and field. This may be taken as the best yield of the past 5 years or, when production conditions are extremely variable such as in dryland farming, 1.5 times the long-term average yield. Once the yield goal has been set, the S requirement is determined directly from relationships such as those illustrated in Table 9-1.

When the soil test available S is greater than the S requirement, the obvious recommendation is that no S be added. If the soil test S is less than the S requirement, an S deficiency is indicated. Depending on the magnitude of deficiency and other aspects of the crop growing conditions, S fertilizer may or may not be recommended. For example, if the deficiency is only 1 or 2 kg/ha, it may be incidentally supplied by rainfall, irrigation, or high rates of a S-containing phosphate fertilizer. However, if the deficiency is much larger than what could realistically be expected to be incidentally supplied, then an S fertilizer or S-containing soil ammendment (such as gypsum) should be applied. When this outcome is encountered, farmers may be well advised to apply 40 to 50 kg S/ha in the form of gypsum as a corrective measure. Except in very permeable soils under high rainfall (low S content), this rate may be sufficient to meet crop needs for several years.

SUMMARY

Although S is usually classified as a secondary nutrient, and as such is required in moderate amounts by plants, widespread deficiencies in crops are uncommon. The chemistry of S favors its conservation in soils either as the sparingly soluble sulfate salts or as a component of organic matter. Thus, recycling plays an important role in supplying crops with S. This, coupled with significant additions from rainfall and irrigation waters ensures adequate S for crops in most environments. Consequently, S soil testing has not received a great deal of attention. Most university soil testing labs do not offer a soil test.

The most common soil S extractant is a weak solution of calcium phosphate from which analysis is made turbidometrically using Ba. Since the available form is mobile in soils, it is appropriate that the subsoil also be tested for S and that the test value be interpreted in relation to crop yield. The total S requirement can be easily calculated from yield goals, taking advantage of already established N-yield goal relationships and the relatively constant ratio of N/S in plants when both nutrients are present in adequate but not excessive amounts.

REFERENCES

Bremer, J.E. 1975. Plant response to sulfur applications on sulfur deficient soil. Ph.D. diss. Oklahoma State Univ., Stillwater (Diss. Abstr. 76-09634).

Cate, R.B., Jr., and L.A. Nelson. 1965. a rapid method for correlation of soil test analyses with plant response data. Int. Soil Testing Series Tech. Bull. no. 1. North Carolina State University, Raleigh.

Council for Agricultural Science and Technology (CAST). 1984. Acid precipitation in relation to agriculture, forestry, and aquatic biology. Report no. 100. Council for Agricultural Science and Technoloby, Ames, IA.

Dijkshoorn, W., and J.E.M. Lampe. 1960. A method of diagnosing the sulfur nutrition status of herbage. Plant Soil 13:227–241.

Eik, K. 1980. Recommended chemical soil test procedures for the Northcentral region. North Dakota Agric. Exp. Stn. Bull. 499.

Freney, J.R., K. Spencer, and M.B. Jones. 1977. On the constancy of the ratio of nitrogen to sulfur in the protein of subterranean clover tops. Commun. Soil Sci. Plant Anal. 8:241–249.

Hue, N.V., F. Adams, and C.E. Evans. 1984. Plant-available sulfur as measured by soil-solution sulfate and phosphate-extractable sulfate in a Ultisol. Agron. J. 76:726–730.

Johnson, C.M., and H. Nishita. 1952. Microestimation of sulfur in plant materials, soils, and irrigation waters. Anal. Chem. 24:736–742.

Johnson, G.V. 1985. Sulfur requirement of Oklahoma crops. Oklahoma Cooperative Extension Facts no. 2237.

Johnson, G.V., and B.B. Tucker. 1982. OSU soil test calibration. Oklahoma Cooperative Extension Facts no. 2225.

Martin, W.E., and J.E. Matocha. 1973. Plant analysis as an aid in the fertilization of forage crops. p. 393–426. *In* L.M. Walsh and J.D. Beaton (ed.) Soil testing and plant analysis. Soil Science Society of America, Madison, WI.

Nuttall, W.F. 1985. Effect of N, P, and S on alfalfa grown on three soil types in northeastern Saskatchewan. II. Nitrogen, P, and S uptake and concentration in herbage. Agron. J. 77:224–228.

Pumphrey, F.V., and D.P. Moore. 1965. Diagnosing sulfur deficiency of alfalfa (*Medicago sativa* L.) from plant analysis. Agron. J. 57:364–366.

Randall, P.J., and K. Spenser. 1980. Sulfur content of plant material: A comparison of methods of oxidation prior to determination. Commun Soil Sci. Plant Anal. 11:257–266.

Reneau, R.B., Jr. 1983. Corn response to sulfur application in Coastal Plain soils. Agron. J. 75:1036–1040.

Sharpley, A.N., S.J. Smith, R.G. Menzel, and R.L. Westerman. 1985. The chemical composition of rainfall in the Southern Plains and its impact on soil and water quality. Oklahoma Agric. Exp. Stn. Tech. Bull. T-162.

Spencer, K. 1978. Sulfur nutrition of clover: Effects of plant age on the composition-yield relationship. Commun. Soil Sci. Plant Anal. 9:883–895.

Stewart, B.A., and L.K. Porter. 1969. Nitrogen–sulfur relationships in wheat (*Triticum aestivum* L.), corn (*Zea mays*), and beans (*Phasedus vulgaris*). Agron. J. 61:267–271.

Westerman, R.L., R.J. O'Hanlon, G.L. Fox, and D.L. Minter. 1983. Nitrogen fertilizer efficiency in bermudagrass production. Soil Sci. Soc. Am. J. 47:810–817.

10 Micronutrient Soil Tests: Correlation and Calibration[1]

F. R. Cox[2]

Several review chapters concerning micronutrient soil tests were published about a decade ago. One appeared in *Micronutrients in Agriculture* (Cox & Kamprath, 1972), two in *Soil Testing and Plant Analysis* (Viets & Lindsay, 1973; Reisenauer et al., 1973), and another in *Soil Testing: Correlating and Interpreting the Analytical Results* (Mortvedt, 1977). More recently, there was another review of the same type except that it concentrated on micronutrient soil testing in the tropics (Lindsay & Cox, 1985). These presentations cover in detail soil reactions, types of extractants, and methodology involved in micronutrient soil tests. Many of the subjects discussed continue to be valid and will not be covered here.

The objective of this chapter will be to review a few selected pieces of research from the past decade. These studies were chosen because they appear to make a unique contribution to our knowledge, and they may indicate the direction for future research. Some comparisons will be made, naturally, with the results of prior studies. A second objective will be to show the present status of micronutrient soil tests in the USA. This will include extractants, interpretations, and recommendations. With this information, it is hoped that a realistic prognosis may be made of research needs for the decade to come.

The current topic is related to one portion of a soil testing program. The phases of a program have been listed in various ways, but a rather classic one may be as follows:

1. Sampling
2. Extractant selection
3. Interpretation
4. Recommendation
5. Promotion

Interpretation is based on correlation and calibration studies. Although this review will concentrate on correlation and calibration experiments, their results are naturally based on certain extractants. Any modification or change

[1]Paper no. 10236 of the Journal Series of the North Carolina Agricultural Research Service, Raleigh, NC 27695-7601.
[2]Professor of Soil Science, North Carolina State University, Raleigh, NC 27695-7619.

in an extractant affects the next several phases in a soil testing program, so some observations will be made first on extractant selection.

EXTRACTANT SELECTION

Extractants are selected based on theoretical and/or practical considerations. Years ago, Bray (1948) outlined three criteria for a good soil test as follows:

1. The extractant should remove all or a proportionate part of the available forms of a nutrient from soils with variable properties.
2. The amount of nutrient extracted should be measured with reasonable accuracy.
3. The amount of nutrient extracted should be correlated with the growth and response of each crop under varying soil conditions.

It is not likely that all three of these conditions will ever be fulfilled completely. The first is rather ambiguous, but it is the foundation of extractant selection based on theoretical considerations. The second is very important for routine laboratories running large numbers of samples. The third is essential, but correlation is a matter of degree and the conditions for which an extractant is applicable may be restricted.

The best recent example of an extractant being selected based on theoretical considerations is the DTPA (diethylenetriaminepentaacetic acid) soil test developed by Lindsay and Norvell (1978). It was developed to extract the micronutrient cations from calcareous soils, so the chelate selection and pH were established specifically for this purpose. The primary micronutrient deficiencies in that region, zinc (Zn) and iron (Fe), were used for correlation and calibration. It may not be an especially practical extractant, however, for two reasons. First, it requires a 2-hour shaking period, more than is desirable for routine laboratory use. Second, its use is restricted to the four micronutrient cations: Zn, copper (Cu), manganese (Mn), and Fe.

Soltanpour and Schwab (1977) modified the DTPA test by shortening the shaking time, combining it with bicarbonate for phosphorus (P) extraction and with ammonium (NH_4) for extraction of other cations. Their modification is now being used and evaluated for the simultaneous extraction of macro- and micronutrients in alkaline soils, and it may become one of the favorite extractants for such soils.

The acid extractants have been used extensively on acid soils. One of the most common is the Mehlich 1 (Mehlich, 1953). This has previously been termed the *double acid*, the *North Carolina extractant*, or by its composition, *0.05 N HCl + 0.025 N H_2SO_4*. It is used extensively for the usual nutrients plus Mn and Zn in the southeastern USA. Recently, another acid extractant has been developed, termed the *Mehlich 3* (Mehlich, 1984). It is proposed to be applicable over a broader range of soil conditions, so it need not be restricted specifically to acid soils. Also, it is an acetate base at pH 2.5 and includes EDTA (ethylenediaminetetraacetic acid) to enhance extraction of certain micronutrients, especially Cu.

CORRELATION AND CALIBRATION

After considering extractants, or in order to compare or evaluate extractants, the next phase in a soil testing program is usually correlation. In this step, the plant concentration or uptake of a nutrient is correlated with the quantity extracted from a group of soils. This is usually done in the greenhouse so that a range of soil conditions may be used. The range of soil conditions may be limited, for instance to calcareous or acid soils, if the extractant should be theoretically restricted within certain bounds. If it is desired to expand that range, or if the range is not known for an extractant that appears to have practical potential, other soil conditions may have to be considered in addition to the amount of micronutrient extracted from the soil. As an example, prior to the advent of chelates for use in alkaline soils, dilute HCl was used to extract Zn. It was known that the presence of free calcium carbonate ($CaCO_3$) reduced the effectiveness of the acid as an extractant, so a soil property related to the presence of free carbonates was included in the correlation. The soil condition that most improved the correlation was "titratable alkalinity" (Nelson et al., 1959).

The idea that soil properties other than the extractable nutrient should be considered when correlating with plant concentration is thus not new, but it is one that has been overlooked in many studies. Unfortunately, there are numerous examples in the literature of comparisons of several extractants that were made without considering any other soil properties.

A Global Study

Recently, there was an extensive study conducted in which soil conditions were evaluated in some detail. It was a global study on the micronutrient status of soils jointly sponsored by FAO and the Institute of Soil Science in Finland (Sillanpaa, 1982). Cooperation was obtained from over 30 nations, with over 3500 sites sampled. These sites were in corn (*Zea mays*) or wheat (*Triticum aestivum*), and both plant material and topsoil samples were sent to the laboratory in Finland for analysis. The soils were extracted for boron (B) and molybdenum (Mo) with single extractants; hot water, and Tamm's reagent, which is ammonium oxalate–oxalic acid (AO–OA), respectively. The micronutrient cations, Cu, Fe, Mn, and Zn, were extracted with two solutions. The first was acid ammonium acetate plus EDTA (AAAc–EDTA) at pH 4.65 (Lakanen & Ervio, 1971) and the second was DTPA at pH 7.3 (Lindsay & Norvell, 1978). Six other soil properties were also determined: soil pH ($CaCl_2$), organic carbon (C), cation exchange capacity (CEC), texture, conductivity, and $CaCO_3$ equivalent.

There was an attempt to correlate the concentration of each nutrient in each crop with the amount extractable and other soil properties. Though some general relations existed, there was too much variability in the plant material to allow such a comparison. Plant samples were taken from numerous varieties and at all stages of growth, and these and other factors added too much variation to the data.

Table 10-1. Correlation between the nutrient concentration in wheat and that
extracted from the soil either alone or after correction for
another soil property (Sillanpaa, 1982).

Soil nutrient	Extraction method	Wheat concentration	
		Alone	Plus correction
		r	
B	Hot water	0.74	0.83
Cu	AAAc-EDTA	0.66	0.73
	DTPA	0.52	--
Fe	AAAc-EDTA	0.33	
	DTPA	0.26	
Mn	AAAc-EDTA	0.04	0.59
	DTPA	0.55	0.71
Mo	AO-OA	0.25	0.70
Zn	AAAc-EDTA	0.67	0.71
	DTPA	0.73	

A single variety of wheat was then grown in the greenhouse. The soils
were fertilized with the macronutrients and magnesium (Mg), but they were
not limed. At harvest, these plants were analyzed for micronutrients. The
correlation between plant and soil nutrient concentrations was much better
for these samples than it was with the field samples (Table 10-1). The correlation coefficients averaged about 0.5 and all were highly significant except the AAAc-EDTA extractable Mn. These relationships accounted for
only about 25% of the variation, however, and it was obvious that they must
be improved to be meaningful. The ratio of plant to soil nutrient concentration was then related to the six other soil properties measured. This was often
significant, indicating that these conditions should also be considered. The
property most related was then included in the correlation by using it to adjust the level of the soil nutrient. Except for Fe, the coefficients were improved markedly with this correction, to about 0.7 in most cases. The first
relationship was poor for Fe, and it was not improved by inclusion of any
other soil condition.

The correction factors for each of the nutrients are shown in Fig. 10-1.
For B, the dominant factor is CEC. The effect of soil B on plant B is more
marked at low CEC than at high CEC. For Cu, the dominant effect is organic
C. As organic C increases, the effect of soil Cu on plant concentration
decreases. Soil pH is the dominant effect for Mn, Zn, and Mo. Soil Mn and
Zn are most effective in increasing the plant content at low pH, whereas soil
Mo is most effective in increasing the plant content at high pH.

In general, including one soil property, in addition to the extractable
nutrient concentration, accounted for the majority of improvement that was
possible in the correlation. There was considerable intercorrelation among
the six soil properties, and in some cases a slight, further improvement could
be made by including another property. For instance, the correlation for soil
Mo alone was 0.25, for soil Mo corrected for soil pH it was 0.70, and for

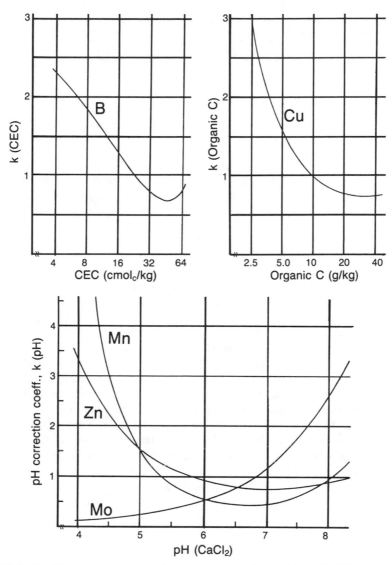

Fig. 10-1. Coefficients for correcting hot water extractable B, AAAc–EDTA extractable Cu and Zn, DTPA extractable Mn, and AO–OA extractable Mo (Sillanpaa, 1982).

soil Mo corrected for both pH and texture it was 0.74. The minor contribution of a second correction was not considered worthwhile and was not utilized.

The objective of this global study was to go a step further and use the data to predict the micronutrient status of the soils in a large part of the world. Whatever its purpose, the approach provides a clear example of the importance of considering other soil properties or conditions in improving

Table 10-2. Relationship between plant B, extractable B by two methods, and other
soil properties in a greenhouse study (Gestring & Soltanpour, 1984).

Extractant	Other properties	Coefficient of determination
		r^2
AB–DTPA	--	0.46
AB–DTPA	Soil pH, OM†	0.70
Hot water	--	0.91
Hot water	Soil pH, OM	0.94

† OM = organic matter.

the relationship between plant and soil concentrations in a correlation study. As stated by Sillanpaa: "These adjustments should be considered as an essential part of soil analysis."

Boron

Another example of incorporating additional soil properties in a soil test interpretation was presented by Gestring and Soltanpour (1984). They compared ammonium bicarbonate–DTPA (AB–DTPA) and hot water as extractants to estimate B availability for alfalfa (*Medicago sativa*) grown on six soils in the greenhouse. The soils were treated with 0, 2, 6, 10, and 20 kg B/ha. Three harvests were analyzed, and the results among the three were similar. Although the correlation between plant and AB–DTPA extractable B was improved markedly by including soil pH and organic matter, the final relationship was still not as good as that with hot water only. Including other properties with hot water improved the relationship very little. The data on the relationships from the first harvest are shown in Table 10-2.

Most people responsible for routine soil tests want to eliminate separate, specialized extractions such as that by hot water. Ponnamperuma et al. (1981) evaluated the use of 0.05 M HCl for B in flooded rice (*Oryza sativa*). This extractant had previously been shown to be useful for Zn. There was a wide range in available B in the 53 soils used, and the quantities extracted by the dilute acid and hot water correlated well (Fig. 10-2). Furthermore, the plant concentration correlated slightly better with the dilute acid ($r = 0.91$) than with the hot water ($r = 0.84$) method. Boron toxicity symptoms were evident on all plants if >4 mg B/kg was extracted with the dilute acid.

Boron deficiency was not considered in the above study, as it is not as common as toxicity in the region. It should be noted, however, that dilute acid extracted only about half as much B as hot water. At low soil concentrations, this reduction would amplify analytical problems. The same would be true in the study discussed previously, which used AB–DTPA. For six of the soils that had low B, the untreated soils had B concentrations of 0.03 and 0.14 mg/kg when extracted with AB–DTPA and hot water, respectively.

The time of refluxing in the hot water procedure is 5 min. In a study of the kinetics of hot water soluble B, Odum (1980) recommended that the time be increased to 10 min. He found the amount extracted to be consistent if the refluxing time was between 10 and 30 min.

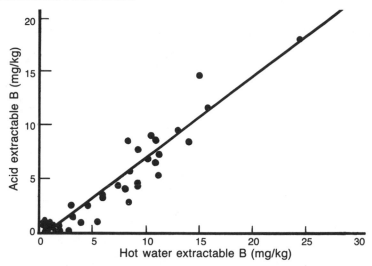

Fig. 10-2. Relationship between 0.05 M HCl and hot water extractable B, where $Y = 0.55 + 0.74X$, $r = 0.96$ (Ponnamperuma et al., 1981).

Mahler et al. (1984) also recommended extending the heating time from 5 to 7 min. Instead of refluxing, however, and dealing with low-B glassware, they utilized sealed plastic pouches. The 7-min heating in pouches gave concentrations similar to refluxing for 5 min.

Copper

Relatively little research has been conducted to interpret soil tests for Cu, especially field calibration. Tills and Alloway (1983) evaluated Cu uptake by wheat growth in the greenhouse on 20 English soils that averaged pH 6.0. Several extractants of both acid and chelate forms were used. Plant Cu correlated best with soil Cu extracted with 0.05 M EDTA at pH 7.0 (Table 10-3). Acid or DTPA extractants were not as good, and AB–DTPA extractable Cu was not significantly correlated with plant Cu. Inclusion of Mn and

Table 10-3. Correlation between the Cu concentration of wheat grown in the greenhouse on 20 soils from England and soil Cu extracted by several methods and on soil Cu plus other soil properties (Tills & Alloway, 1983).

	Plant Cu	
Extractant	r^2	R^2†
AB–0.005 M DTPA (pH 7.6)	0.14	0.47
0.005 M DTPA (pH 7.3)	0.49	0.64
0.05 M EDTA (pH 7.0)	0.75	0.79
Dilute HNO₃	0.50	0.59
1.0 M HCl	0.45	0.59

† Mn-oxide, Fe-oxide, clay, pH, and loss on ignition included.

Table 10-4. Correlation between percent maximum yield of several crops grown in the greenhouse and soil Cu extracted by four methods and the critical level derived from the relationship (Makarim & Cox, 1983).

Extractant	Percent yield	Critical level
	R^2†	mg/dm³
	Greenhouse ($n = 15$), three crops	
AB–DTPA	0.47	0.53
Mehlich 1	0.70	0.26
Mehlich 3	0.73	0.37‡
Mehlich-Bowling§	0.60	0.62
	Field ($n = 7$), two crops	
Mehlich-Bowling	--	0.70¶

† Linear plateau.   § Mehlich and Bowling, 1975.
‡ Wrist action shaker.   ¶ Cate and Nelson, 1965.

Fe oxides, clay, pH, and loss on ignition increased the accounted for variation, but EDTA alone provided the most practical interpretation.

The apparent poor showing of AB–DTPA in the study cited above may be associated with differences in sample handling or extracting conditions. Numerous workers have shown that many factors; grinding, glassware, shaking, temperature, etc., may affect the quantity of a micronutrient extracted (Shuman, 1980; Soltanpour et al., 1976, 1979; Lindsay & Norvell, 1978). The effect of shaker type and associated glassware was noted in a study of four extractants for Cu in North Carolina (Makarim & Cox, 1983). With the Mehlich 3 extractant, there was about 25% less Cu extracted with a wrist action than with a platform shaker. This study involved 15 acid soils, and the results with AB–DTPA were not quite as good as those with acid extractants (Table 10-4).

Grundon and Best (1982) evaluated the response of wheat to DTPA extractable Cu on 35 sites in Australia. They found yield responses and/or low grain Cu (<3 mg/kg) if the DTPA extractable Cu was <0.4 mg/kg on gray or gray-brown soils. For red or red-brown soils, however, DTPA Cu concentration as low as 0.2 mg/kg did not indicate a deficient condition. Furthermore, there was no relationship between yield response and soil pH, P, or Zn concentration. Although soil class may be used to improve a soil test interpretation, it does not identify the causal relationship, nor is soil class a continuous function.

Iron

Attempts to develop soil tests for Fe have not been highly successful. As indicated previously, the DTPA method of Lindsay and Norvell (1978) had this objective and an excellent theoretical background. Havlin and Soltanpour (1981) evaluated a modification of this technique (AB–DTPA) that is currently used in Colorado. Sorghum [*Sorghum bicolor* (L.) Moench] was grown in the greenhouse on 40 soils. A critical level of 4.8 mg Fe/kg was established. Deficient and nondeficient categories were separated just as well with AB–DTPA as with DTPA alone.

Other workers have been less successful in interpreting Fe soil tests. Katyal and Sharma (1984) used chelates, both DTPA and EDTA, on 16 soils used to grow upland rice in the greenhouse. They found no relationship between percent yield and extractable soil Fe. The authors also evaluated the effects of soil pH, lime, organic C, and clay levels and found them unrelated to percent yield, either singly or together. In fact, the only consistent factor related to percent yield was the ferrous-Fe concentration of the plant.

Some interesting research is now being conducted on Fe reduction and uptake. Organic compounds are currently being identified that may be associated with this process (Mino et al., 1983; Ripperger & Schrieber, 1982). When the factors related to the production of these compounds in certain plant species are known, it should be possible to incorporate them into an Fe soil test interpretation. It may be that the concentration of other micronutrients, or the ratio among micronutrients, may be an important consideration.

Manganese

Manganese deficiency is probably the most important micronutrient problem in the Atlantic Coastal Plain region of the USA. It occurs when the soils are limed at substantially more than the recommended rate. Soil pH is, therefore, an integral part of the Mn soil test interpretation.

Recently, there have been two extensive field studies in the southeastern USA conducted primarily on Ultisols to calibrate the Mn soil test for the cultivars of soybean [*Glycine max* (L.) Merr.] currently being grown. In experiments conducted in North Carolina, two extractants, Mehlich 1 (M1) and Mehlich 3 (M3) were used on 38 site-year observations (Mascagni & Cox, 1985). Five cultivars of soybean were grown and yield response to applied Mn measured. Yield response was evaluated in terms of extractable Mn and soil pH by multiple regression. When the predicted yield response is zero, the soil Mn and pH combinations of critical level may be shown, as in Fig. 10–3. As the soil pH is increased from 6 to 7, for example, the concentration of M3 extractable Mn (M3Mn) at which there is no yield response to applied Mn increases from 4 to 8 mg/dm^3 (volume basis equivalent of mg/kg). As soil pH becomes higher or the soil Mn lower than these combinations, the yield response to Mn fertilization increases.

This is a rather difficult concept to explain because yield response to Mn is dependent on two soil parameters. An availability index was developed, therefore, to pool these effects. In North Carolina, an index value of 25 is synonymous with the critical level and positive responses to fertilization are expected at any lower values. As shown in Fig. 10–4, the yield response is clearly inversely related to the availability index, M3MnAI. This makes the Mn soil test interpretation much more simple and direct.

A similar group of experiments was conducted in Virginia during this same period (Gettier et al., 1985). Eight soybean cultivars were grown on

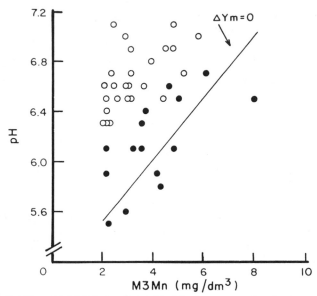

Fig. 10-3. Soil pH and Mehlich 3 extractable Mn for responsive (open symbols) and nonresponsive (closed symbols) sites for soybean production. Any point on the ΔY_m = 0 line is a prediction of the "critical level" conditions (Mascagni & Cox, 1985).

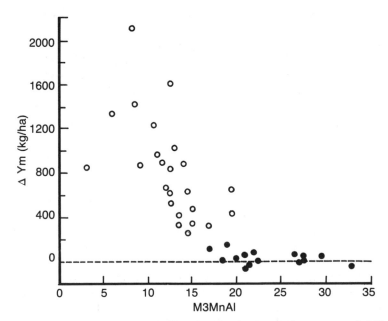

Fig. 10-4. Yield response of soybean to Mn fertilization in relation to an availability index derived from both soil pH and M3 extractable Mn as follows: M3MnAI = 101.7 − 15.2 pH + 3.75 M3Mn (Mascagni & Cox, 1985).

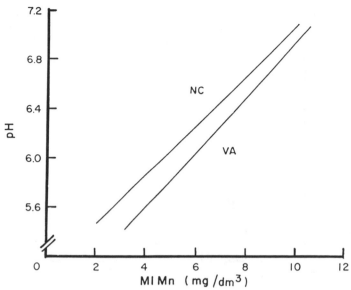

Fig. 10-5. Prediction of the "critical level" conditions for soybean from North Carolina (Mascagni & Cox, 1985) and Virginia (Gettier et al., 1985) studies in relation to soil pH and Mehlich 1 extractable Mn.

more than 30 site-year observations. Mehlich 1 extractable Mn (M1Mn) was used and yield response was expressed as a percent of the maximum. Again, both soil Mn and pH had to be included to arrive at a meaningful interpretation, and the critical level combination of these parameters was depicted at an expression of 90% of maximum yield. Their data were converted to a volume basis by assuming a volume weight of 1.25 g/cm^3, and when the pH is between 6 and 7, the M1Mn critical level is slightly greater than that found in the North Carolina study (Fig. 10-5), but the overall relationships are similar.

The effect of pH on the Mn soil test interpretation with acid extractants appears to be the same with crops less sensitive to Mn deficiency. Mascagni and Cox (1984) found that the pH effect on the estimate of the Mn critical level for soybean and corn was essentially identical (Fig. 10-6). For corn, a crop less sensitive to Mn deficiency than soybean, the critical level of soil Mn was at least 2 mg/dm^3 less than it was for the soybean.

Soil pH has also been shown to be a dominant factor in interpreting chelate extractions for Mn. Sheppard and Bates (1982) grew barley (*Hordeum vulgare*) in the greenhouse on about 70 Canadian soils and evaluated seven extractants plus other soil properties. Manganese uptake was used as the dependent variable. Manganese extracted by DTPA was not a valid parameter, but the concentration removed by AB-DTPA or an acid such as phosphoric acid (H_3PO_4) entered readily into the interpretation.

Sheppard and Bates then applied the results of the greenhouse experiment to observations made at 11 field locations, four of which were Mn deficient for winter wheat or soybean. The greenhouse predictions proved valid on the coarse textured soils, but not on the fine-textured ones.

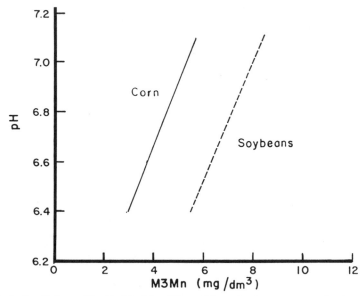

Fig. 10-6. Prediction of the "critical level" conditions for two crops varying in sensitivity to Mn deficiency in relation to soil pH and Mehlich 3 extractable Mn (Mascagni & Cox, 1984).

Zinc

Other soil properties have also been considered in the Zn soil test interpretation. These properties are often intercorrelated. It is difficult to differentiate between effects of certain properties such as texture and CEC. In a regional study that included the diagnosis of Zn problems in corn production, CEC appeared to be more important than clay (Cox & Wear, 1977). The interpretation was restricted to soils with a CEC of <7.5 cmol$_c$/kg. It is apparent from the responses to Zn found at the two sites shown in Fig. 10-7 that the prediction equation was not valid for the site with a CEC of 14 cmol$_c$/kg.

In this regional study, two acid extractants and a chelate were compared, and there was little difference among them. It was also of no benefit to include soil pH in the interpretation. Although the effect of acidity on Zn uptake is not expected to be as great as that for Mn, soil pH has been found to be useful when interpreting Zn soil tests. Haq and Miller (1972) studied the Zn concentration in corn grown in the greenhouse on 85 soils. Soil Zn from three chelate and one acid extractant were used along with other soil properties. The two best extractants, EDTA and DTPA, gave coefficients of determination with plant Zn of about 0.37 when used alone, but when the pH linear and quadratic terms were included this was increased to 0.75. The relationship between plant Zn and these two soil properties is shown in Fig. 10-8. The effect of pH is most apparent on the acid soils. A similar response has been shown with DTPA (Bates, 1984).

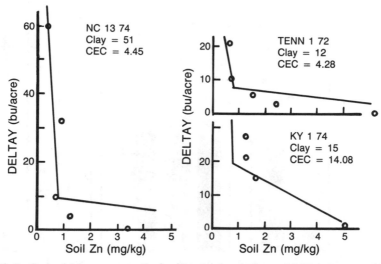

Fig. 10-7. Corn yield response to Zn fertilization in relation to Mehlich 1 extractable Zn at several clay and CEC levels (taken from Cox & Wear, 1977).

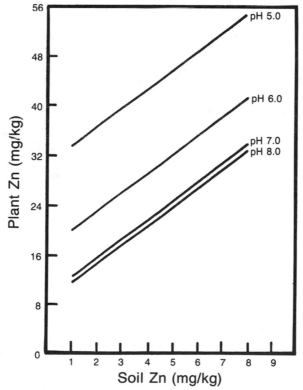

Fig. 10-8. Concentration of Zn in corn in relation to EDTA extractable Zn and soil pH (Haq & Miller, 1972).

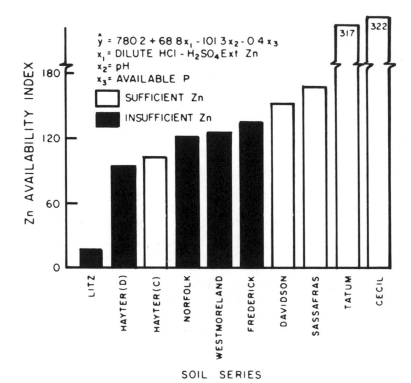

Fig. 10-9. The Zn availability index predicted by a three-variable regression equation for 10 soils that supplied either sufficient or insufficient Zn for corn growth (Alley et al., 1972).

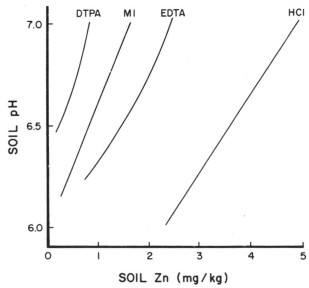

Fig. 10-10. Prediction of the "critical level" conditions for corn in relation to soil pH and extractable Zn by the method indicated. Data from sources cited in text.

It has also been shown that pH should be considered when interpreting a Zn soil test utilizing an acid extractant. Alley et al. (1972) used the Mehlich 1 (M1) extractant to assay Zn and P and determined the relationship between Zn uptake by corn grown in the greenhouse and soil Zn, pH, and P. They then used this relationship to identify responsive field sites (Fig. 10-9), and the results were favorable. Although three parameters are involved, all are readily available from a routine soil test.

Peaslee (1980) also used a multiple regression approach with 0.1 N HCl extractable Zn, soil pH, and Bray 1 extractable P. The results identifying critical conditions for this and the three studies previously mentioned are shown in Fig..10-10. The similarity in the slopes indicate a uniform effect of pH among extractants.

Molybdenum

There have been very few efforts to correlate or calibrate a Mo soil test. The Grigg or Tamm reagent, acid ammonium oxalate, remains the primary extractant used. Soil pH, however, is a more important factor than soil Mo in determining Mo uptake or yield response. This was shown in a greenhouse study with cauliflower (*Brassica oleracea botrytis*) by Karimian and Cox (1979) and in regional field studies on soybean (Anderson & Mortvedt, 1982) and forages (Mortvedt & Anderson, 1982).

STATUS OF MICRONUTRIENT SOIL TESTS

The second objective to be covered in this presentation is to show the current status of micronutrient soil tests in the USA. In 1983, Carl Gray conducted a survey of the State Soil Testing Laboratories for the Soil Testing and Plant Analysis Committee (S877) of the Soil Science Society of America. The methodology for each micronutrient was given by state, and the extractant, soil/solution ratio, time, and technique of measurement were listed. This survey was updated in 1985 and additional information was requested on interpretations and recommendations. For each micronutrient the following were given; (i) primary crop(s) showing a deficiency, (ii) interpretation or critical level, (iii) interacting factors or limitations, and (iv) the recommendation.

The details on critical levels and recommended rates are quite clear, but the extent to which the states consider the interacting factors or how they affect the interpretation is not. In many cases an interacting factor is known to exist, but apparently does not affect the critical level; in other cases it is a factor that must be considered. Examples of such will be given in the following discussion, but only a summary of the interpretation for the current usage of each micronutrient soil test will be presented.

Hot water extractable B is the singular direct measure of B availability in use at present (Table 10-5). Twenty states employ this method, and the primary crop used for interpretation is alfalfa, which was cited as most im-

Table 10-5. Soil test extractants for B, interacting factors considered in the
interpretation, crops for which recommendations are made,
and critical level data used in the USA.

Extractant	Interacting factors	Crops	Critical level	
			Mean	Range
			mg/dm³ or mg/kg	
Hot water (20)†	Crop (5)	Alfalfa (16)	0.7	0.1–2
	pH (6)	Beet‡ (2)		
	H₂O (4)	Sugar beet‡ (2)		
	Soil type	Peanut† (2)		
	OM (2)	Cotton (2)		
	Yield goal	Corn (4)		
	Lime history	Vegetables (6)		
None	pH >6	Cotton	--	--

† Numbers in parentheses indicate the number of times reported by states; no number
infers unity.
‡ Beet and Sugar beet, *Beta vulgaris;* Peanut, *Arachis hypogaea* L.

portant in 16 states. Over all crops, the average critical level was 0.7 mg/dm³
and the range was 0.1 to 2 mg/dm³. (Volume measures will be assumed
although some laboratories are weighing samples.)

One of the dominant interacting factors with B is crop, as there has been
considerable research and observations on the relative susceptibility of crops
to B deficiency. Five states listed crop as an interacting factor. Factors listed
next were pH, H₂O, soil type, and organic matter. Soil type would encom-
pass texture, the factor shown most important by Sillanpaa (1982) in the
global study.

One state also listed a soil property used for interpretation, namely pH,
even though no B analysis was conducted on the soil. One might speculate
on several reasons why this is done, such as a difficult analysis, a sensitive
crop (cotton, *Gossypium hirsutum* L.) and an economical fertilizer material.

The second most common micronutrient assay by a single method is Zn
by DTPA (Lindsay & Norvell, 1978). Nineteen states are conducting this
analysis; 15 have interpretations for corn and nine for bean (*Phaseolus
vulgaris* L.) (Table 10-6). The average critical level is 0.8 mg/dm³ and the
range is 0.25 to 2 mg/dm³. Lime, meaning high lime, calcareous, or high
bicarbonate, was cited as the most important interacting factor by four states.
There was no indication, however, that this factor was used to adjust the
critical level. Soil pH was listed by only one state as an important factor.

The next two extractants most commonly used for Zn were Mehlich 1
(eight states) and 0.1 *M* HCl (six states). Soil pH and P were listed in about
half of these states as factors affecting interpretation, and in two of the states
the interpretation was directly dependent on soil Zn, pH, and P.

The extractant most commonly used for Mn is Mehlich 1 (Table 10-7).
Eight states use this extractant; all make the interpretation for soybean, and
all include pH as an integral part of the interpretation. The critical level varies
from 4 mg/dm³ at pH 6 to 8 mg/dm³ at pH 7.

Table 10-6. Soil test extractants for Zn, interacting factors considered in the
interpretation, crops for which recommendations are made,
and critical level data used in the USA.

Extractant	Interacting factors	Crops	Critical level	
			Mean	Range
			mg/dm³ or mg/kg	
DTPA (19)†	Lime (4)	Corn (15)	0.8	0.25-2
	P (2)	Bean‡ (9)		
	pH	Sorghum (4)		
	OM	Rice (3)		
	Irr.	Flax‡ (3)		
		Fruit trees (3)		
Mehlich 1 (8)	pH (5)	Corn (7)	1.1	0.5-3
	P (3)	Pecan‡ (2)		
0.1 M HCl (6)	pH (3)	Corn (5)	5	2-10
Others§ (5)				

† Numbers in parentheses indicate the number of times reported by states; no number
infers unity.
‡ Bean, *Phaseolus vulgaris* L.; flax, *Linum usitatissium* L.; pecan, *Carya illinoensis*
(Wangenh.) K. Koch.
§ Mehlich 3, NH₄HCO₃ + DTPA, pH 3 Ac (acetate), and pH 4.8 Ac.

Table 10-7. Soil test extractants for Mn, interacting factors considered in the
interpretation, crops for which recommendations are made,
and critical level data used in the USA.

Extractant	Interacting factors	Crops	Critical level	
			Mean	Range
			mg/dm³ or mg/kg	
Mehlich 1 (8)†	pH (8)	Soybean (8)	7	4 at pH 6 to 8 at pH 7
DTPA (5)	--	Small grain (3) Corn (3)	1.4	1-2
0.03 M H₃PO₃ (3)	pH (3) OM (2)	Oat‡ (2) Soybean (3) Oat (2)	10 10	0-20 0-20
Others§ (3)				

† Numbers in parentheses indicate the number of times reported by states; no number
infers unity.
‡ *Avena sativa* L.
§ Mehlich 3, 0.1 M HCl, and no Mn analysis.

The next most important extractants for Mn were DTPA (five states)
and 0.03 M H₃PO₄ (three states). The states using H₃PO₄ also cited pH as
an important factor in interpretation.

Iron availability is highly dependent on extraneous conditions, and, as
such, there is little trust in most Fe analyses. In fact, the largest number of
states (eight) indicated that Fe interpretations were made solely on the follow-

Table 10-8. Soil test extractants for Fe, interacting factors considered in the interpretation, crops for which recommendations are made, and critical level data used in the USA.

Extractant	Interacting factors	Crops	Critical level	
			Mean	Range
			mg/dm³ or mg/kg	
No Fe analysis (8)†	Soil pH or calcareous	Soybean (5)	--	--
	Aeration, H₂O (3)	Sorghum (3)	--	--
	Calcareous and symptoms (2)			
DTPA (7)	Lime	Sorghum (3)	4	2.5-5
		Fruit trees (3)		
Others‡ (3)				

† Numbers in parentheses indicate the number of times reported by states; no number infers unity.
‡ NH₄HCO₃ + DTPA, Mehlich 1.

ing conditions: calcareous or pH (four states), aeration or water (three states), and calcareous *and* symptoms (two states; Table 10-8). There were seven states, however, that did use DTPA extractable Fe at a critical level averaging 4 mg/dm³. The primary crops listed were soybean, sorghum, and fruit trees.

Very few states analyze soils for Cu, and the mean critical levels given in Table 10-9 for DTPA and Mehlich 1 do not seem realistic; they are too high. Organic matter was listed several times as an interacting factor, but there was no indication that it was ever used directly in the interpretation.

There were five states that listed the methodology to analyze soils for Mo. No state, however, used soil Mo concentration for interpretation. Nine states do use acidity, the limit varying from pH <5.5 to <7.0, to evaluate the need for Mo for selected crops. These crops, of course, were primarily legumes.

To summarize the results of this survey; soil tests are conducted most often for three micronutrients: B, Zn, and Mn. Boron is extracted only with

Table 10-9. Soil test extractants for Cu, interacting factors considered in the interpretation, crops for which recommendations are made, and critical level data used in the USA.

Extractant	Interacting factors	Crops	Critical level	
			Mean	Range
			mg/dm³ or mg/kg	
DTPA (4)†	OM	Small grain	0.8	0.12-2.5
	Crop	Corn (2)		
	pH			
Mehlich 1 (2)	OM (2)	--	3	0.1-10
Others‡ (3)				

† Numbers in parentheses indicate the number of times reported by states; no number infers unity.
‡ 1.0 *M* HCl, Mehlich 3, no Cu analysis.

hot water, and no other soil properties are considered to affect the critical level. Zinc is extracted with DTPA, primarily on calcareous soils, and with dilute acids on acidic soils. There is an important soil acidity effect on the interpretation, and pH and soil P concentration are now being considered by some states. Soil Mn is extracted with acids and pH is a dominant factor in the interpretation.

Few states conduct soil tests for Fe because of poor correlation with response or with Cu because of infrequent deficiencies. There are no states that use extractable Mo as a criteria for interpretation.

SUMMARY

Recent research has indicated that properties other than the extractable level of the micronutrient in question should almost always be included in a soil test interpretation. Iron may be the exception, but as further research clarifies the role of the plant under Fe deficient conditions, other soil properties may be important for it too.

The inclusion of other soil properties in interpretation has been accomplished most consistently and completely for Mn; soil pH is an integral part of the interpretation. Acidity is being considered in the Zn interpretation by a few states, and this could be expanded. Research also needs to be conducted on the role of CEC, or texture, on the interpretation of hot water extractable B. And finally, the role of organic matter on Cu availability needs further investigation. Total organic C or organic matter estimated by oxidation may not be the best fraction to express this effect.

REFERENCES

Alley, M.M., D.C. Martens, M.G. Schnappinger, Jr., and G.W. Hawkins. 1972. Field calibration of soil tests for available zinc. Soil Sci. Soc. Am. Proc. 36:621–624.

Anderson, O.E., and J.J. Mortvedt. 1982. Soybeans: Diagnosis and correction of manganese and molybdenum problems. Southern Coop. Ser. Bull. 281. University of Georgia, Athens.

Bates, T.E. 1984. Soil test for zinc availability revised. Progress report. Department of Land Resource Science, University of Guelph, Guelph, Ontario, Canada.

Bray, R.H. 1948. Requirements for successful soil tests. Soil Sci. 66:83–89.

Cate, Jr., R.B., and L.A. Nelson. 1965. A rapid method for correlation of soil test analysis with plant response data. Int. Soil Testing Series Tech. Bull. 1. North Carolina Agric. Exp. Stn., Raleigh.

Cox, F.R., and E.J. Kamprath. 1972. Micronutrient soil tests. p. 289–317. *In* J.J. Mortvedt et al. (ed.) Micronutrients in agriculture. Soil Science Society of America, Madison, WI.

Cox, F.R., and J.I. Wear. 1977. Diagnosis and correction of zinc problems in corn and rice production. Southern Coop. Ser. Bull. 222. North Carolina State University, Raleigh.

Gestring, W.D., and P.N. Soltanpour. 1984. Evaluation of the ammonium bicarbonate-DTPA soil test for assessing boron availability to alfalfa. Soil Sci. Soc. Am. J. 48:96–100.

Gettier, S.W., D.C. Martens, and S.J. Donohue. 1985. Soybean yield response prediction from soil test and tissue manganese levels. Agron. J. 77:63–67.

Grundon, N.J., and E.K. Best. 1982. Survey of the extent of copper deficiency of wheat on the Western Downs, Queensland. Queensland J. Agric. Anim. Sci. 39:41–46.

Haq, A.U., and M.H. Miller. 1972. Prediction of available soil Zn, Cu, and Mn using chemical extractants. Agron. J. 64:779–782.

Havlin, J.L., and P.N. Soltanpour. 1981. Evaluation of the NH_4HCO_3-DTPA soil test for iron and zinc. Soil Sci. Soc. Am. J. 45:70–75.

Karimian, N., and F.R. Cox. 1979. Molybdenum availability as predicted from selected soil chemical properties. Agron. J. 71:63–65.

Katyal, J.C., and B.D. Sharma. 1984. Association of soil properties and soil and plant iron to iron deficiency response in rice (Oryza sativa L.). Commun. Soil Sci. Plant Anal. 15:1065–1081.

Lakanen, E., and R. Ervio. 1971. A comparison of eight extractants for the determination of plant available micronutrients in soils. Suom. Maataloustiet. Seuran Julk. 123:232–233.

Lindsay, W.L., and F.R. Cox. 1985. Micronutrient soil testing for the tropics. In P.L.G. Vlek (ed.) Micronutrients in tropical food crop production. Fert. Res. 7:169–200.

Lindsay, W.L., and W.A. Norvell. 1978. Development of a DTPA soil test for zinc, iron, manganese, and copper. Soil Sci. Soc. Am. J. 42:421–428.

Mahler, R.L., D.V. Naylor, and M.K. Frederickson. 1984. Hot water extraction of boron from soils using sealed plastic pouches. Commun. Soil Sci. Plant Anal. 15:479–492.

Makarim, A.K., and F.R. Cox. 1983. Evaluation of the need for copper with several soil extractants. Agron. J. 10:903–909.

Mascagni, H.J., Jr., and F.R. Cox. 1984. Diagnosis and correction of manganese deficiency in corn. Commun. Soil Sci. Plant Anal. 15:1323–1333.

Mascagni, H.J., Jr., and F.R. Cox. 1985. Calibration of a manganese availability index for soybean test data. Soil Sci. Soc. Am. J. 49:382–386.

Mehlich, A. 1953. Determination of P, K, Na, Ca, Mg, and NH_4. Soil Test Div. Mimeo. North Carolina Department of Agriculture, Raleigh.

Mehlich, A. 1984. Mehlich 3 soil test extractant: A modification of Mehlich 2 extractant. Commun. Soil Sci. Plant Anal. 15:1409–1416.

Mehlich, A., and S.S. Bowling. 1975. Advances in soil test methods for copper by atomic absorption spectrophotometry. Commun. Soil Sci. Plant Anal. 6:113–128.

Mino, Y., T. Ishida, N. Ota, M. Inoue, K. Nomoto, T. Takemoto, H. Tanaka, and Y. Sugiura. 1983. Mugineic acid–iron (III) complex and its structurally analogous cobalt (III) complex: Characterization and implication for absorption and transport of iron in gramineous plants. J. Am. Chem. Soc. 105:4671–4676.

Mortvedt, J.J. 1977. Micronutrient soil test correlations and interpretations. p. 99–117. In T.R. Peck et al. (ed.) Soil testing: Correlating and interpreting the analytical results. Spec. pub. 29. American Society of Agronomy, Madison, WI.

Mortvedt, J.J., and O.E. Anderson. 1982. Forage legumes: Diagnosis and correction of molybdenum and manganese problems. Southern Coop. Ser. Bull. 278. University of Georgia, Athens.

Nelson, J.L., L.C. Boawn, and F.G. Viets. 1959. A method for assessing zinc status of soils using acid-extractable zinc and "titratable alkalinity" values. Soil Sci. 88:275–283.

Odum, J.W. 1980. Kinetics of the hot water soluble boron soil test. Commun. Soil Sci. Plant Anal. 11:759–765.

Peaslee, D.E. 1980. Effect of extractable zinc, phosphorus, and soil pH on zinc concentrations in leaves of field-grown corn. Commun. Soil Sci. Plant Anal. 11:417–425.

Ponnamperuma, F.N., M.T. Cayton, and R.S. Lantin. 1981. Dilute hydrochloric acid as an extractant for available zinc, copper, and boron in rice soils. Plant Soil 61:297–310.

Reisenauer, H.M., L.M. Walsh, and R.G. Hoeft. 1973. Testing soils for sulphur, boron, molybdenum, and chlorine. p. 173–200. In L.M. Walsh and J.D. Beaton (ed.) Soil testing and plant analysis. Soil Science Society of America, Madison, WI.

Ripperger, H., and K. Schrieber. 1982. Nicotinamine and analogous amino acids, endogenous iron carriers in higher plants. Heterocycles 17:447–461.

Sheppard, S.C., and T.E. Bates. 1982. Selection of a soil extraction and a multiple regression model to predict plant available manganese. Commun. Soil Sci. Plant Anal. 13:1095-1111.

Shuman, L.M. 1980. Effects of soil temperature, moisture, and air drying on extractable manganese, iron, copper, and zinc. Soil Sci. 130:336-343.

Sillanpaa, M. 1982. Micronutrients and the nutrient status of soils: A global study. Food and Agriculture Organization of the United Nations (FAO) Soils Bull. 48. United Nations, Rome.

Soltanpour, P.N., A. Khan, and W.L. Lindsay. 1976. Factors affecting DTPA-extractable Zn, Fe, Mn, and Cu from soils. Commun. Soil Sci. Plant Anal. 7:797-821.

Soltanpour, P.N., A. Khan, and A.P. Schwab. 1979. Effect of grinding variables on the NH_4HCO_3-DTPA soil test values of Fe, Zn, Mn, Cu, P, and K. Commun. Soil Sci. Plant Anal. 10:903-909.

Soltanpour, P.N., and A.P. Schwab. 1977. A new soil test for simultaneous extraction of macro- and micronutrients in alkaline soils. Commun. Soil Sci. Plant Anal. 3:195-207.

Tills, A.R., and B.J. Alloway. 1983. An appraisal of currently used soil tests for available copper with reference to deficiencies in English soils. J. Sci. Food Agric. 34:1190-1196.

Viets, F.G., Jr., and W.L. Lindsay. 1973. Testing soils for zinc, copper, manganese, and iron. p. 153-172. In L.M. Walsh and J.D. Beaton (ed.) Soil testing and plant analysis. Soil Science Society of America, Madison, WI.

11 Field Experimentation: Changing to Meet Current and Future Needs[1]

Malcolm E. Sumner[2]

Before attempting to arrive at the conclusion implicit in the title of this paper, it is necessary to begin with an historical evaluation of what has and is being done in this field. There are a number of papers dealing with this topic, some of which are cited here for the interested reader (Collis-George & Davey, 1960; Cope & Evans, 1985; Hanway, 1967, 1973). The approach generally adopted has been succinctly summarized by Hanway (1973):

> To develop a quantitative relationship between different measured levels of any one component and the yields obtained, it is necessary to conduct experiments with different measured levels of that factor and to measure the resultant yields. If the levels of other components. . .influence the relationship it is necessary either (i) to hold the levels of all other components constant and to develop the relationship at this "standard" level of the other components, or (ii) to conduct additional studies to determine the effects of the other components. . .and interactions between different components. Where more than one factor influences yield and interactions exist between the different variables and where levels of the different factors can be controlled, a factorial experimental design is used to obtain the data needed to determine the desired relationships. Where levels of a given factor cannot be controlled, experiments may be conducted at different measured levels of that factor as they occur naturally and then a relationship can be developed by means of regression.
>
> In all instances, the levels of factors affecting yield must be measured. If the level of these factors can be expressed numerically, we can quantitatively describe the system. . .If the results from different experiments and experimenters are to be combined to develop the desired relationships, common (standard) methods of measuring each variable in the systems must be developed and used. Suitable standards for each variable in the systems must be developed, generally accepted, and used if field experimentation is to develop into the science it can be.

We shall now proceed to analyze the present situation in the hope of being able to shed light on the path we should follow in the future.

---

[1] Contribution from Department of Agronomy, College Station, University of Georgia, Athens, GA 30602.

[2] Professor of Agronomy, University of Georgia, Athens, GA 30602.

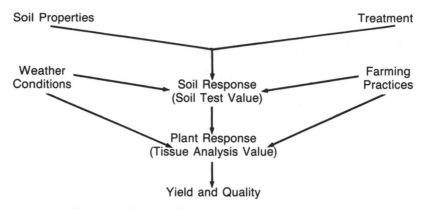

Fig. 11-1. Factors affecting treatment-yield relationship.

FIELD EXPERIMENTS

The designs of field experiments, used most frequently in soil fertility research aimed at providing data for soil test calibration, have been of the type involving analysis of variance procedures such as randomized block, Latin square, and factorial designs. These statistical designs basically ask the question: Is treatment A different from treatment B and, in certain cases, do interactions exist? Thus, they measure the relationship between treatments such as levels of a nutrient and the corresponding yields. This relationship is schematically illustrated in Fig. 11-1 with the intermediate steps involved.

Analysis of variance simply measures the relationship between treatment and yield, ignoring reactions that may take place between the nutrient added and the soil constituents, resulting in a decrease or increase in the quantity of the nutrient available to the root for uptake. Thus, a nutrient treatment usually reacts with the soil to give a soil response that can be measured by an appropriate soil testing procedure. As an illustration, the soil response to 50 kg P/ha would be entirely different on an Oxisol than on a Mollisol of similar texture, with the crop reacting in an entirely different manner. Soil response to a treatment, in addition, depends on weather conditions and farming practices. For example, a topdressing of ammonium-nitrate (NH_4-NO_3) constitutes no treatment and therefore brings about no soil response without adequate rainfall or irrigation to dissolve the material so that it enters the root zone. Similarly, a surface application of limestone causes little soil response unless it is incorporated. Thus, it is meaningless to say treatment A is better than treatment B, as would be measured by analysis of variance, unless we also measure the corresponding soil test levels of the particular nutrient. However, even if we measure the soil test levels, they cannot be analyzed by analysis of variance procedures, particularly when factorial experiments are used, because the treatments often result in different soil responses arising from interactions with other treatments.

To further emphasize the disregard with which the importance of the above intermediate steps in the treatment-yield relationship has met, a recently published bulletin dealing with experimental designs for agrotechnology transfer (Silva, 1981) makes only one fleeting reference to the importance of measuring soil response to a treatment. It then continues to state:

> To date, such analyses, especially in the case of nitrogen, have not proven ac- curate enough: none of the sampling procedures employed, the extracting solu- tions used, the soil extracting procedures utilized, or the precision of the chemical determinations have been able to duplicate the capacity of a plant to obtain its nutrients from the soil. (Capo, 1981).

Although there is some truth in the above quotation, it goes much too far in suggesting only a poor relationship, at best, between soil tests and yield. There is a substantial body of evidence in the literature to the contrary (Walsh & Beaton, 1973)!

One of the assumptions implicit in the use of field experiments is that the site is reasonably uniform and if it is not, it is assumed that a uniformity trial was carried out prior to the commencement of the experiment in order to measure and compensate for any fertility anomalies. Modern day pressures usually preclude the latter option, which often results in erroneous conclu- sions being drawn.

A further problem associated with but not implicit in classical field ex- perimentation has been that of site selection. There are many cases, both published and unpublished, where experiments have been performed on a site without care being taken to ensure that the treatments selected are likely to result in a yield response. Without the potential of a yield response, such experimentation is quite useless for the purpose of soil test calibration. In addition, many soil test calibration studies have been carried out under less than optimal conditions where uncontrollable factors, such as rainfall, have dominated nutrient responses. Under such conditions, it is very difficult, if not impossible, to measure yield responses that could very well be obtained if some of these uncontrollable factors were more favorable. This then points to the need for soil test calibrations to be conducted at the highest possible yield levels, which would minimize the adverse effects of the uncontrollable factors. How this may be achieved will be discussed later in the paper.

Additional problems arise from the fact that such experimental designs limit the number of levels and the number of nutrients that can be studied simultaneously. In nonfactorial designs, one is limited to a number of levels of a single or combination of nutrient(s) with all other nutrient factors being held constant at so-called "optimum" levels. If nutrient balance is impor- tant in crop production, and there is a large body of evidence (Dibb & Thomp- son, 1985; Follett et al., 1981; Munson, 1968; Sumner, 1978; Sumner & Farina, 1986) that says that it is, then it is impossible for nutrients held con- stant to be at levels appropriately balanced for each of the levels of the nutrient being varied. Thus the "constant optimum" levels of background nutrients introduce bias into the experiment by favoring one particular level of the nutrient being varied. Although some of these problems can be over- come by the use of factorial designs involving levels of multiple factors, the ideal experiment that might involve five levels of the 13 essential elements cannot be conducted as it would require an area of the order of 12 million ha, quite clearly an impossible task! Even factorials with fewer factors and levels exhibit the same problems that are involved in the single factor approach. Partial factorial designs introduce some improvement but it would seem that the ultimate goal of determining optimally balanced levels for all essential nutrients cannot be attained using traditional experimental designs.

Let us for a moment look at the benefits farmers would like to obtain from soil testing. The types of questions they ask are: What are the soil test levels of the essential elements required to obtain optimum yields? How much fertilizer must I apply to raise my soil test from its current level to the optimum? Is the optimum soil test level of one nutrient influenced by the soil test level of another, and if so, how can I know that the levels of all the nutrients in my soil are at their appropriate and balanced levels? Do the cur-

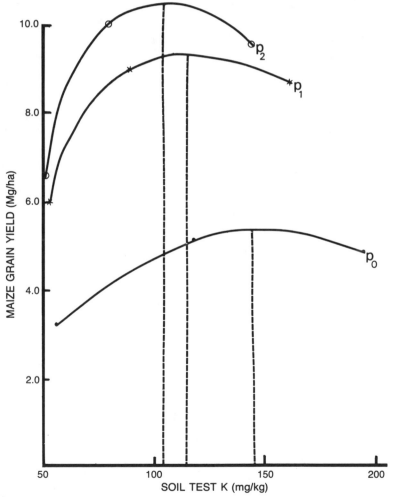

Fig. 11-2. Relationship between soil test K and maize grain yield at three levels of soil P on an Oxisol (Sumner, unpublished results).

rently used calibrations obtained at yield levels lower than those of the present day apply to my production system, and if not, what can be done? We need to discuss the extent to which the classical field experimental approach answers these questions as they are the ones that should receive our attention as agronomists.

Classical field experiments do not answer the first question (what are the soil test levels for the essential elements required to obtain optimum yields?) without the introduced bias of assuming that nutrients not being varied are at optimum levels and by interactions of the controlled nutrients with uncontrollable environmental factors from season to season. Such problems are illustrated in the following examples. In Fig. 11-2, the effects of P levels on the optimum soil test level for K are illustrated using data from a phosphorus × potassium (P × K) factorial conducted by the author on an Oxisol. It is quite clear that the level of added P affects the optimum level of soil test K. Thus, soil test calibrations derived from single factor experiments are likely to suffer some bias because the preselected level of a background or basal nutrient would tend to preferentially favor one of the levels of the nutrient being varied. The effects of seasonal variations in environmental factors on the relationship between soybean [*Glycine max* (L.) Merr.] yield and soil test P (Hargrove et al., 1984) is illustrated in Fig. 11-3. As the authors pointed out, it was virtually impossible to arrive at an optimum soil test P value from these data because seasonal variations obvious-

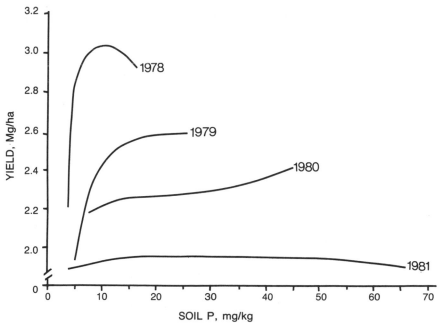

Fig. 11-3. Effect of season on the relationship between soil test P and soybean yield (after Hargrove et al., 1984).

ly were much greater than the treatment effects with which they interacted, resulting in quite variable responses between years.

The second question (How much fertilizer must I apply to raise my soil test from its current level to the optimum?) is addressed almost perfectly by field experiments and requires little further elaboration. All that is necessary to arrive at the answer is to plot the levels of a nutrient applied against soil test values. The slope of the line gives the amount required to raise the soil test value by one unit.

Factorial experiments offer some insight into the third question (Is the optimum soil test level of one nutrient influenced by the soil test level of another and if so, how can I know that the levels of all the nutrients in my soil are at their appropriate and balanced levels?), but fail to produce a complete answer because of the limited number of levels and factors that can be dealt with simultaneously. It is therefore doubtful whether one can state with any degree of certainty that the optimum soil test values derived from limited factorials are, in fact, in optimal balance and are correct.

In terms of the final question (Do the currently used calibrations obtained at yield levels lower than those of the present day apply to my production system, and if not, what can be done?), currently used calibrations are probably inadequate, particularly if they were derived from data obtained at low yield levels. However, since it is not the topic of the present paper, this subject will not be discussed further.

As mentioned above, the relationship between soil response or soil test level and yield is not readily studied using analysis of variance techniques. Many workers have tackled this problem by using simple or multiple regression techniques, the appropriateness of which will now be explored.

REGRESSION

Many of the problems associated with the attempts to relate soil test values to yields obtained from field experiments have been solved, to some extent, by the use of regression techniques.

Regression may simply be defined as the process by which the empirical functionality between dependent and independent variables is established. It usually involves the fitting of some equation to the data relating a dependent variable to one or more independent variables, the first case being an example of simple and the latter multiple regression. In order to evaluate the usefulness of this approach in describing the relationship between yield (dependent variable) and soil test values (independent variables), we need to investigate the assumptions that are made when fitting such relationships.

First, we should be congnizant of the fact that there are numerous factors that govern the yield of a crop, only a few of which can be controlled or measured in a meaningful manner. Thus, when we attempt to relate soil test values for a single nutrient to yield, the first assumption that we make is that *all* the other factors governing yield are either at their "optimal, standard, or mean" levels or that they are such that they are not primarily governing yield in this particular case. The difference between the R^2 value and 1.0 indicates the extent of unaccounted variation due to factors not being measured in the model. It can be readily seen that this assumption results in a vary simplistic model and the relationship derived is unlikely to be applicable in situations differing very much from which it was obtained.

The data presented in Fig. 11-2 illustrate that the optimal soil test level of one nutrient depends on the levels of the others. Thus, a single factor regression approach can lead to an optimum soil test value for a particular nutrient; however, this optimum is unlikely to be valid at different levels of other nutrients. Multiple regression analysis provides a method for determining the effect of one variable at different levels of other factors being changed. This approach is a definite improvement over the single factor approach but both of the relationships obtained suffer from the disadvantage of always being season- and frequently site-specific. Furthermore, this approach does not quite fully answer the question in which farmers are primarily interested. What they are seeking is the optimum soil test level for all the important nutrients and not the optimum level of one nutrient at some standard or mean level of the others. Thus, even if we are able to obtain a multiple regression equation that accounts for 99% of the variation in yield for a particular set of data from a soil test calibration experiment (which is highly improbable!), it would be totally useless in predicting the relationship the following year because we have no way of knowing how the uncontrollable factors will present themselves. Thus, regression techniques might not be the ultimate answer in soil test calibration problems, but the shape of the curves obtained permits one to estimate optima for soil test parameters that have proven to be useful in diagnostic work.

PROPOSALS FOR THE FUTURE

As a general conclusion from the foregoing discussion, one can state that analysis of variance techniques have been of little benefit in soil test calibration work save for providing data suitable for manipulation by regression techniques, which have provided us with calibrations that have been useful in diagnostic work. Because regression equations for soil test calibrations are both site and season specific, however, one has been forced to repeat such work both in space and time in order to represent a range of the uncontrollable variables.

In order to make further progress in this area, one must first build a conceptual model of the system for which we wish to obtain soil test calibrations. To that end, let us attempt to conceptualize the most general model of the relationship between a soil test parameter and yield. This can be readily done by a series of questions and answers.

1. Is it possible to obtain a high yield when the value of the soil test parameter is very low? Answer: No, because it is likely to limit yield by being present in an insufficient amount.

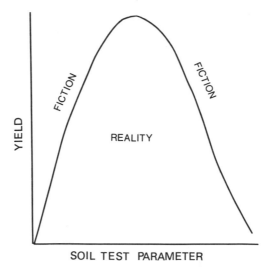

Fig. 11-4. Conceptual model of the relationship between a soil test parameter and yield.

2. Is it possible to obtain a low yield when the value of the soil test parameter is low? Answer: Yes, because there is probably sufficient nutrient to allow some growth and a low yield.

3. Is it possible to obtain a high yield when the value of the soil test parameter is optimum? Answer: Yes, provided all other factors are at optimal or near optimal levels.

4. Is it possible to obtain a low yield when the value of the soil test is optimum? Answer: Yes when one or more other factors severely limit the yield.

5. Is it possible to obtain a high yield when the value of the soil test parameter is very high? Answer: Probably not, because it is likely to be toxic or out of balance with the other nutrients.

6. Is it possible to obtain a low yield when the value of the soil test parameter is very high? Answer: Yes, because the nutrient is unlikely to be so toxic or out of balance to preclude any yield.

This conceptual model is depicted in Fig. 11-4 and we shall proceed to use it as a basis for the continued discussion.

The curve drawn in Fig. 11-4 is a boundary line (Beaufils, 1971; Webb, 1972; Walworth et al., 1986) separating an area labeled "reality" in which any observation validly made would lie from that area labeled "fiction" where valid observations would *never* be found. This boundary line thus represents the general limiting case for the relationship between yield and a soil test parameter. By continually adding to a data bank and plotting the particular relationship, one would constantly challenge this boundary line so that ultimately, provided sufficient data are available, its precise location could be established. This would then represent the ultimate soil test calibration. The value of the soil test parameter where the maximum yield is obtained

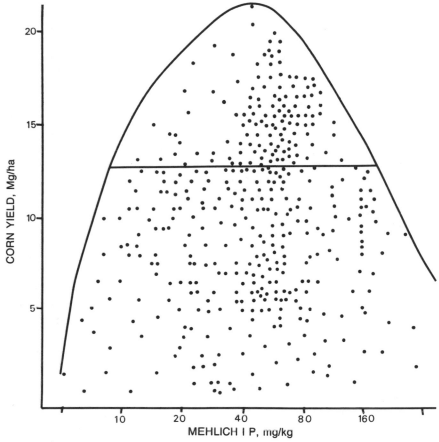

Fig. 11–5. Scatter diagram together with confining boundary line for the relationship between soil test P and corn yield on Coastal Plain soils of the southeastern USA.

represents the level of the particular nutrient that would be optimal under any conditions. By interpolation, one could establish the maximum yield that might be obtained at any particular soil test level or alternatively, establish the minimum soil test level for a given required yield.

An actual data set consisting of soil test P measured by Mehlich I extractant on soils of the southeastern USA and corn (*Zea mays* L.) yield is presented in Fig. 11–5 to illustrate the scatter of points obtained in a real world situation. The data used in this example were derived from observations made in farmers' fields (Sumner, unpublished data) as well as from field experiments (Sumner, unpublished data; Woodward, 1985) over a number of years. It can be seen that most points occur in the center of the plot around the value for soil test P where the boundary line peaks. As one moves closer and closer to the boundary line the density of points diminishes, because for a point to lie on the boundary line requires that the levels of all other growth factors be at or very near the optimum, thereby reducing the frequency of such observations. Looked at another way, a soil test value near the boundary line will provide sufficient of that nutrient to obtain the corresponding yield, provided the levels of the other growth factors are optimal. When one or more of the other growth factors are limiting, this yield will not be obtained and it is not possible to predict the degree of yield reduction. If one imposes a yield ceiling, it is then possible to arrive at a range of soil test values that would be adequate for any particular yield. This is illustrated by the horizontal line in Fig. 11–5, which shows that for a corn yield of 12.5 Mg/ha, the soil test P level can lie between 10 to 160 mg/kg.

In collecting data for use in plots such as that in Fig. 11–5, it is extremely important that the soils be analyzed by the same appropriate method and the variability between soils in terms of mineralogy and chemistry be kept to a minimum. This is necessary in order to obtain the most sensitive calibrations. Furthermore, it is important that the soils be sampled at the equivalent time, preferably after fertilizer applications have been made so that the quantities of nutrients available to the crop are reflected in the soil test value. Neglect of these precautions will lead to unnecessary variability with a resultant lack of diagnostic precision.

In the past, there have been many cases where workers have forced regression models on their data that are entirely inappropriate but have been used because preconceived models so dictated. Such data, taken from Hanson (1979), are presented in Fig. 11–6 as an illustration. He used a segmented line model to fit his data and stated: "This model indicates a linear relationship up to soil available P of 8.83 μg g^{-1} and relative yield of 93.2% with no increase when soil available P is above this level." However, the highest recorded yield point (113%) occurs at a soil test value of 18 mg/kg. The 20% yield increase, claimed to be of little consequence above, translates into an actual yield difference of about 0.8 Mg soybean grain/ha, worth at least U.S. $150 at 1986 prices. To ignore such differences, even if statistics so dictate, seems to be ludicrous. The data in Fig. 11–6 appear to lend themselves better to an interpretation using the boundary line approach as indicated by the dotted line confining the points. Fitting of regression equations to such data assumes that yields, corresponding to the points such as those in circles, are being determined primarily by the level of soil P, which is probably untrue. On the other hand, this would more likely be true of the points encompassed in squares. Equal treatment of these two sets of points in the regression analysis seems to bear little relationship to what happens in the real world.

In terms of future work on soil test calibration, it would seem that the use of the boundary line technique might prove to be rewarding because it overcomes many of the problems associated with the classical field experiments and regression techniques enumerated above. However, this does not mean that we should abandon field experiments, because they have an appropriate and important role to play in solving these problems. For example, any type of field experiment that is capable of generating variability supplies particularly useful information to data banks for use in boundary line interpretations. Appropriate experimental designs in this catagory would be

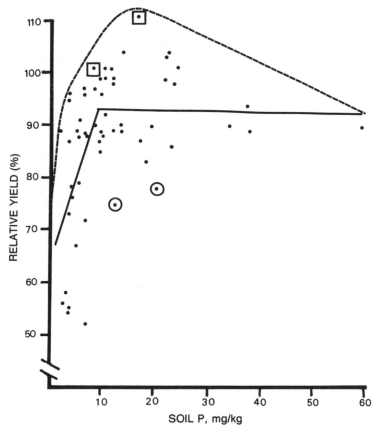

Fig. 11-6. Segmented line model fit for relationship between Bray-P and relative yield (after Hanson, 1979). Dotted line is a boundary line inserted by the author that confines the data.

complete and partial factorials, preferably not replicated, but rather used as single replications at a number of sites and continuous function experiments (Fox, 1973) that are analogous to two-way interaction tables laid out in the field without randomization. In both cases, it is important to select sites likely to be responsive to the treatments applied. Despite the fact that these are the only designs capable of generating substantial multifactor variability in the field, results from other experimental designs used for research in soil fertility and other facets of agronomy such as variety tests can also be used as sources for data banks. Therefore, any validly obtained observation consisting of yield and soil test value can be used in the boundary line technique. In this regard, data obtained from surveys about farmers' fields are particularly useful because they are likely to represent most of the types of variability occurring in the real world for which a calibration is required.

As a challenge, this author proposes that we set about collecting soil data obtained by a particular method together with yields so that data banks can be built up from which improved calibrations can be obtained. Data already existing in the files of agronomists as well as newly collected data are suitable for this purpose. Data banks so developed would be available for use in soil test calibration or other research.

The above considerations deal largely with formal research approaches to soil test calibration. However, we frequently lose sight of the fact that the best calibrations for a farmer's field would be obtained from that particular field. Despite the obviousness of this statement, we pay little or no attention to its importance. At present, many soil test laboratories analyze countless millions of soil samples without ever obtaining any feedback from the producer as to what kind of success or failure was achieved. Herein lies an enormous potential data base. If the producer could be encouraged or coerced into supplying yield and other production data to the soil test laboratory, large meaningful data banks could be cheaply established. It has been stated that this would be impossible to achieve, but surely there is considerable potential to be tapped. Even if such an attempt failed, there would still be the possibility of using soil test data from an individual field for calibration purposes. The farmer, together with his county agent or consultant, can use successive years' data to plot the course of soil fertility in the field and on that basis fine tune the recommendations. Soon most progressive farmers will have their own in-house microcomputers, which are admirably suited for this purpose. This type of data storage and manipulation will result in improved fertilizer usage and other management practices.

In conclusion, it can be said that we often suffer from "hardening of the categories" in that we are reluctant to change our ways for one reason or another. Change is definitely indicated in the way we conduct field experimentation for soil test purposes!

SUMMARY

Traditionally, field experiments used for soil fertility research have comprised designs that are analyzed by analysis of variance procedures. Critical analysis of this approach shows that it best serves the elucidation of a treatment-yield relationship, which is not the relationship sought in soil testing. In soil testing one is interested in the relationship between soil test value and yield, which cannot be analyzed by analysis of variance. Regression is better suited to quantifying the latter relationship but also has some limitations such as uncontrolled variation in many of the parameters affecting the relationship. A proposal for using a boundary line technique to interpret the soil test-yield relationship is made as well as recommendations for the type of field experimentation to be conducted for this purpose.

REFERENCES

Beaufils, E.R. 1971. Physiological diagnosis—A guide for improving maize production based on principles developed for rubber trees. FSSA J. 1:1.

Capo, B.G. 1981. Consideration on experimental design and production functions. p. 159–166. *In* J.A. Silva (ed.) Experimental designs for predicting crop productivity with environmental and economic inputs for agrotechnology transfer. Dep. Pap. 49. Hawaii Institute of Tropical Agriculture and Human Resources, University of Hawaii, Honolulu.

Collis-George, N., and B.G. Davey. 1960. The doubtful utility of present-day field experimentation and other determinations involving soil-plant relationships. Soil Fert. 23:307–310.

Cope, J.T., and C.E. Evans. 1985. Soil testing. p. 201–224. *In* B.A. Stewart (ed.) Advances in soil science I. Springer-Verlag, New York.

Dibb, D.W., and W.R. Thompson, Jr. 1985. Interactions of potassium with other nutrients. p. 515–533. *In* R.D. Munson (ed.) Potassium in agriculture. American Society of Agronomy, Crop Science Society of American, and Soil Science Society of America, Madison, WI.

Follett, R.H., L.S. Murphy, and R.L. Donahue. 1981. Fertilizers and soil amendments. Prentice-Hall, Inc., Englewood Cliffs, NJ.

Fox, R.L. 1973. Agronomic investigations using continuous function experimental designs—Nitrogen fertilization of sweet corn. Agron. J. 65:454–456.

Hanson, R.G. 1979. Effect upon soybean cultivar Bragg, when P is band-concentrated upon variable soil-available P. (sic). Agron. J. 71:267–271.

Hanway, J.J. 1967. Field experiments for soil test correlation and calibration. p. 103–114. *In* G.W. Hardy (ed.) Soil testing and plant analysis, Part I. Spec. Pub. 2. Soil Science Society of America, Madison, WI.

Hanway, J.J. 1973. Experimental methods for correlating and calibrating soil tests. p. 55–66. *In* L.M. Walsh and J.D. Beaton (ed.) Soil testing and plant analysis. Soil Science Society of America, Madison, WI.

Hargrove, W.L., F.C. Boswell, and J.T. Touchton. 1984. Correlation of extractable soil phosphorus and plant phosphorus with crop yields for double cropped wheat and soybeans. Univ. of Georgia Res. Bull. 304. University of Georgia, Athens.

Munson, R.D. 1968. Interaction of potassium and other ions. p. 321–353. *In* V.J. Kilmer et al. (ed.) The role of potassium in agriculture. American Society of Agronomy, Crop Science Society of America, and Soil Science Society of America, Madison, WI.

Silva, J.A. 1981. Experimental designs for predicting crop productivity with environmental and economic inputs for agrotechnology transfer. Dep. Pap. 49. Hawaii Institute of Tropical Agriculture and Human Resources, University of Hawaii, Honolulu.

Sumner, M.E. 1978. Interpretation of nutrient ratios in plant tissue. Commun. Soil Sci. Plant Anal. 9:335–345.

Sumner, M.E., and M.P.W. Farina. 1986. Phosphorus interactions with other nutrients and lime in field cropping systems. p. 201–236. *In* R. Stewart (ed.) Advances in soil science, Vol. 5. Springer-Verlag, New York.

Walsh, L.M., and J.D. Beaton (ed.). 1973. Soil testing and plant analysis. Soil Science Society of America, Madison, WI.

Walworth, J.L., W.S. Letzsch, and M.E. Sumner. 1986. Use of boundary lines in establishing diagnostic norms. Soil Sci. Soc. Am. J. 50:123–128.

Webb, R.A. 1972. Use of the boundary line in the analysis of biological data. J. Hort. Sci. 47:309–319.

Woodard, H.J. 1985. Nutrient assimilation and yield response of grain corn (*Zea mays* L.) under various fertilizer and soil moisture regimes and high plant populations. Ph.D. diss. Rutgers University, New Brunswick, NJ (Diss. Abstr. 85–20419).

12 The Value and Use of Soil Test Summaries[1]

S. J. Donohue[2]

Since the 1940s, when soil testing became recognized as a reliable agronomic tool, millions of soil samples have been analyzed by soil testing laboratories in the USA. Today, approximately 3 million samples per year are analyzed by various laboratories around the nation. Many of these laboratories prepare summaries of soil test results. In a survey conducted in 1985 of the 50 states, 46 land grant university or government laboratories offered a soil testing service. Of these 46, 40 prepared summaries periodically, i.e., once every 1 to 10 years or so. Most of them, 25, produced summaries every year. These summaries are used to evaluate the fertility status of the soil to determine general fertilizer and lime needs. They are also used in farmer/homeowner educational programs to bring about desired changes in fertilizer and lime use.

VALUE AND RELIABILITY OF SOIL TEST SUMMARIES

Questions arise, from time to time, on the real value and reliability of soil test summaries to represent the true fertility status of the soil. Various factors have been reported (Harrison et al., 1983; McCollum & Nelson, 1954; Parker et al., 1951) that may bias soil test results upward or downward. These factors are discussed in the sections below.

Soil Samples Submitted by Better Farmers

Several workers have stated that most soil samples analyzed by soil testing laboratories have been submitted by better farmers. This would tend to bias test results upward and not give a true picture of the average fertility status of the soil. In the early 1950s, McCollum and Nelson (1954) conducted a study in Duplin County, North Carolina (Table 12–1), where about 1200 systematically collected samples were compared with over 500 farmer-collected samples that were tested by the North Carolina Department of

[1] Contribution of the Department of Agronomy, Virginia Polytechnic Institute and State University, Blacksburg, VA 24061.

[2] Professor and Extension Agronomist, Department of Agronomy, Virginia Polytechnic Institute and State University, Blacksburg, VA 24061.

Table 12-1. Average soil test values from Duplin County, North Carolina
(adapted from McCollum & Nelson, 1954).

Source of samples	Soil pH	P	K	Organic matter
		kg/ha		g/kg
Systematic	5.3	112	61	16
Farmer	5.5	94	80	18

Table 12-2. Average soil test values from six counties in north central Wisconsin.†

Source of samples	pH^2	Lime requirement	P	K	Organic matter
		Mg/ha	— kg/ha —		g/kg
Survey	6.05	6.5	80	217	37
Summary	6.08	6.5	83	209	34

† Adapted from Peters, 1978.

Agriculture Soil Testing Laboratory. In this study, test results were fairly close except for potassium (K). It is important to note that soil test values were averaged over 10 crops; with K, large differences were found with the truck crops between systematically collected samples and farmer-submitted samples. This caused much of the difference in K level for the farmer samples. These researchers concluded that even though differences occurred, they were not of sufficient magnitude to invalidate a soil test summary of farmer-submitted samples.

A more recent study, conducted by Peters in Wisconsin (Peters, 1978, Table 12-2), provides information regarding biases in soil test results from only the better farmers submitting soil samples. The primary objective in Peter's study was to evaluate the lime requirement of soils in the north central Wisconsin drift region, which were developed over acid glacial till, in order to determine future lime needs. For this research, 2600 soil samples were randomly collected throughout the six-county region. The soil test summary from the University of Wisconsin Soil Testing Laboratory was not used initially because it was felt that primarily, better farmers submitted samples, and the soil test summary would not show the true lime needs for this region. After the random samples were collected and summarized, they were compared with the soil test summary for 1974 through 1977 (73 000 samples) from the Wisconsin Soil Testing Laboratory. Very little difference was found between the two separately collected samples. Peters concluded that there was very little bias in soil test summaries generated by soil testing laboratories.

Inclusion of Troubleshooting Samples

Another factor that has been mentioned as possibly effecting the accuracy of soil test summaries is the inclusion of troubleshooting samples, or those samples submitted from problem fields, in soil test summaries. This would tend to bias the test results downward. However, troubleshooting samples normally comprise only a small percentage of the total samples received by a laboratory. A comparison was made of the total number of soil samples submitted to the Virginia Tech Soil Testing and Plant Analysis Laboratory in June and July, when troubleshooting normally occurs in Virginia, with the total number of samples processed by the laboratory per year, from 1980 through 1984. An average of only 7.5% (range = 7.2 to 7.9% of a total annual average sample number of 109 334 samples) of the total samples was received during these 2 months each year. Of these samples, <5%, probably 1 to 2%, were actually troubleshooting samples. Thus total troubleshooting samples accounted for only a small portion of the total samples received and therefore this is most likely not a great problem.

Summarizing over Physiographic Regions

A third factor that may affect accuracy of soil test summaries is summarizing over physiographic regions. Workers have felt that this technique, summarizing over different physiographic regions with different soil properties, will bias results. A study was conducted in the late 1970s by researchers in Kentucky (Harrison et al., 1983) where random samples collected in two different counties were compared to farmer samples submitted to the Kentucky Soil Testing Laboratory (Table 12-3). In Graves County, with primarily loess-derived soils, soil test results between the randomly collected vs. farmer-submitted samples were similar. In Todd County, with different physiographic regions, results differed, most noticeably with phosphorus (P) and K. When samples were separated according to physiographic region, the reason for the differences became apparent. In southern Todd County, most soils were derived from limestone, and the land was gently rolling. In northern Todd County, the soils were derived primarily from sandstone and shale with some limestone, and the topography was hilly to steeply sloping. This large variation in soils accounted for the difference in test results in Todd County between the two sets of samples.

Table 12-3. Average soil test values from two Kentucky counties (adapted from Harrison et al., 1983).

Source of samples	Average soil test			
	Water pH	Buffer pH	P	K
			—— kg/ha ——	
		Graves County		
Randomly taken	6.10	6.59	40	251
Sent by producers	6.14	6.55	40	243
		Todd County		
Randomly taken	6.37	6.60	65	361
Sent by producers	6.46	6.58	47	311
Random, southern	6.48	6.66	81	436
Random, northern	6.25	6.55	46	273

Table 12–4. Potassium soil test levels for conventionally tilled corn grown for grain in Virginia, 1984 (Donohue, 1985).

Soil test level	Coastal Plain		Piedmont		
	Southern	Northern	Southern	Northern	Mountain
	% of samples				
Low + medium	92	80	74	68	64
High + very high	7	20	26	32	36

A comparison was made of soil samples submitted from the three major physiographic regions in Virginia to the Virginia Tech Soil Testing and Plant Analysis Laboratory in 1984 (Table 12–4). Considerable differences were found in K test results, due primarily to the different soils in the regions. In general, the soil texture varied from loamy sands to sandy loams between the southern and northern Coastal Plain regions; to sandy loams and sandy clay loams, with the clay being primarily of the kaolinitic type, in the Piedmont region; and then to loams to clay loams, with clays of both kaolinitic and montmorillonitic types, in the mountain region. The K-holding ability of the soil, being closely related to soil texture and type of clay mineral, is illustrated rather graphically in Table 12–4. Therefore, in preparing and using summaries to most accurately reflect the true fertility status of the soil, it is best to separate the data into physiographic regions. If a summary of soil test results is desired for counties with two or more physiographic regions, the county could perhaps be separated into physiographic regions using zip code or post office box number information from the field history sheets submitted with the soil samples.

Summarizing over Crops

A fourth factor influencing the accuracy of soil test summaries is summarizing over crops. Samples submitted for crops that are fertilized and/or limed differently would be expected to have different average soil test levels. Table 12–5 contains a summary for Virginia for 1984 where one physiographic region was selected and five different cropping categories were compared. Considerable differences in soil test results were found between corn and

Table 12–5. Phosphorus soil test levels for five crops grown in the mountain region of Virginia, 1984 (Donohue, 1985).

Crop	No. of samples	P soil test level			
		Low	Medium	High	Very high
		% of samples			
Corn for silage, no-till	1872	1	20	55	24
Native pasture	3014	25	51	20	4
Improved pasture	954	14	48	32	6
Tobacco, burley	564	1	18	45	36
Gardens	2751	2	13	27	58
Weighted avg	9155	11	31	32	27

pasture, between native and improved pasture, and between these crop categories and tobacco (*Nicotiana tabacum* L.) or gardens where fertilizer is usually applied in excessive amounts. Therefore, to obtain the best estimate of the fertility status of the soil, one needs to summarize data for individual crops.

Time of Year Sample Collected

A fifth factor that may affect the accuracy of a soil test summary is the time of year that samples are collected. Soil pH, P, and K levels tend to vary during the year, which can affect soil test results. In a study reported by Liebhardt and Teel (1977), extractable soil K increased gradually from 65 to 81 kg/ha between June through December 1975, to 90 kg/ha in March 1976, and then to 134 kg/ha in May 1976, just prior to planting. Soil K levels then declined during the summer. This study was conducted on a sandy Coastal Plain soil and no fertilizer was applied during the time of the experiment. The fluctuation in K was measured over several growing seasons. In addition to K, soil pH levels also fluctuate, normally being lowest in the fall and highest in the spring. However, if the relative proportion of samples received in the fall and spring are the same from year to year, this should not have a significant influence on soil test summaries. In an evaluation of soil samples sent to the Virginia Tech Soil Testing and Plant Analysis Laboratory over a 5-year period in the late 1970s, fall vs. spring sample numbers remained fairly consistent, with 40 to 45% of the samples submitted in the fall and 55 to 60% of the samples submitted in the spring.

Additional Factors Affecting Summaries

Several other factors have also been mentioned that can effect the accuracy of soil test summaries. These are year-to-year variation in soil test levels, climatic differences within a region, land on short-term lease, collection of a representative sample by the farmer, use of mean vs. median, and differences in soil test methods and calibrations between states in a region.

Regarding year-to-year variation in soil test levels; this has been documented for certain tests such as soil pH, but biases can be minimized by summarizing over 2 or 3 years instead of only 1 year.

In regard to climatic differences within a region; sizeable differences are sometimes found in large counties or physiographic regions and can, indeed, have an effect on test results. In Whitman County, Washington, depending on the location within the county, average annual rainfall will vary from 356 to 559+ mm. Similar climatic differences might be found over a given physiographic region. However, these different areas within a county or physiographic region can be separated (e.g., by zip code, etc.), if desired, and individual summaries prepared.

Soil samples submitted from land on a short-term lease may affect soil test results because land used in this manner will usually receive minimal fertilizer and lime applications; i.e., only that which is needed to grow the current crop. However, the number of samples submitted from leased land is usually small and one can assume, with some accuracy, that the same relative number of samples from leased land is submitted for testing each year.

Regarding collection of representative soil samples; it is understood that not all farmers follow recommended sampling procedures for collecting good representative samples, but this percentage probably does not change much from year to year.

In regard to the use of mean vs. median, if one is averaging soil test results and reporting averages, they can be biased upward by a small proportion of samples that test extremely high. It would probably be better to report the median in this situation. However, since most laboratories that prepare summaries separate their test results into "low," "medium," and "high" numerical ranges, this problem is not considered very serious.

Regarding differences in soil test methods and calibrations between states; this can and does effect the accuracy of soil test summaries that are prepared on a regional (USA) level. Summaries of this type are most often prepared by large fertilizer and lime companies as well as the educational arms of the fertilizer and lime industries. These summaries are rough approximations, at best, of soil nutrient levels and should be interpreted as such. Very few laboratories agree on the actual numerical range for "low," "medium," and "high" for a given nutrient and a given soil test method, although the general concept of crop response at these test levels (Thomas & Peaslee, 1973) is generally accepted. Therefore, this type of summary should be viewed with some caution.

In the questionnaire sent out in the summer of 1985, laboratory representatives were asked to respond to questions concerning biases in soil test summaries. Of the 40 laboratories preparing summaries, 39 responded. Approximately 44% considered biases to be minimal, whereas 56% indicated that biases were considerable, but summaries were still useful in evaluating general trends in soil fertility levels. Soil test summaries, with careful interpretation, can provide useful information on soil nutrient levels and fertilizer and lime needs.

USE OF SOIL TEST SUMMARIES

The two major uses of soil test summaries are (i) to evaluate fertilizer and lime recommendations, and (ii) to encourage the proper use of fertilizer and lime. These two aspects of soil test summary use often follow each other in close order and examples will be given that illustrate this point. Various types of summaries can be prepared to convey information on soil nutrient levels. Soil fertility status can be expressed using "low," "medium," and "high" nutrient ranges or by comparing the percent needing or not needing fertilizer and lime. One can make single-year comparisons to evaluate the need for fertilizer and lime or make multi-year comparisons to evaluate how needs are changing. Summaries can be made on a county, regional, or state-wide basis. Examples of these different types of summaries are shown below.

Table 12-6. Long-term changes in P soil test level for corn grown in Virginia.

Area	Years	No. of samples	P soil test level			
			Low	Medium	High	Very high
			% of samples			
State	1970, 1971	13 508	7	28	44	22
	1983, 1984	29 572	3	26	59	12
Southampton County,	1970, 1971	1 295	0	6	58	36
Virginia	1983, 1984	4 781	0	8	79	12

Evaluation of P Recommendations

Table 12-6 contains the results of an evaluation in long-term changes in P soil test levels for corn (*Zea mays* L.) grown in Virginia. Prior to 1976, P fertilizer was recommended at the "very high" test level in Virginia in a soil fertility maintenance program. In 1976, with high fertilizer prices and the more important fact that one would not expect a crop response at this level (i.e., no return would be expected on the fertilizer investment), P recommendations were discontinued at "very high." At county producer meetings in the late 1970s, response to P fertilizer at different soil test levels was emphasized. In comparing a 1970 and 1971 (2-years) summary for Virginia (G.W. Hawkins, unpublished data), with a more recent 1983 and 1984 summary (Donohue, 1985), this change in P recommendations was reflected in the decrease in the number of samples testing in the "very high" category. In Southampton County, Virginia's largest agricultural county, a similar pattern was found. The Extension educational program, considered to be responsible for at least part of this decrease, was deemed to be successful.

Evaluation of Soil pH and Lime Needs

A comparison similar to the P evaluation discussed above was made to evaluate soil pH changes over time for Southampton County, Virginia (Table 12-7). In the 1960s, the Virginia Tech Soil Testing and Plant Analysis Laboratory recommended an optimum pH of 5.8 for the corn-peanut (*Arachis hypogaea*) rotation to prevent manganese (Mn) deficiency in peanut. In the ensuing years, however, too many low soil pH problems occurred; farmers were not monitoring their pH levels carefully, and the soil pH sometimes dropped into the range at which aluminum (Al) became toxic,

Table 12-7. Long-term changes in soil pH for corn grown
in Southampton County, Virginia.

Years	No. of samples	Soil pH				
		0.0-4.9	5.0-5.4	5.5-5.9	6.0-6.4	6.5+
		% of samples				
1970, 1971	1295	2	10	48	36	4
1983, 1984	4781	0	3	20	57	20

Table 12-8. Long-term acidification of farmland in six Washington and six northern Idaho Counties (adapted from Mahler et al., 1985).

Soils testing below pH 6.0	Eastern Washington			Northern Idaho		
	1960	1972	1980	1960	1972	1980
%	no. of counties					
<15	12	9	6	12	3	2
15–45	0	2	5	6	4	2
46–65	0	1	0	0	4	2
>65	0	0	1	0	1	6

causing poor growth. A decision was made in the mid-1970s to increase the optimum pH for this rotation to 6.5 and to instruct the producers to routinely spray for Mn deficiency. This is reflected in the increase in number of samples testing >pH 6.0 in 1983 and 1984, as compared to 1970 and 1971.

Soil test summaries were used by workers in Washington and Idaho to document the acidification of the soil in the eastern Washington-northern Idaho area over the last 25 years (Table 12-8). This acidification was brought about by the change in crop rotation from a cereal crop-legume sequence to a mostly continuous wheat (*Triticum aestivum* L.) rotation with acid-forming nitrogen (N) fertilizer as the N source. The number of counties testing < pH 6.0 increased dramatically between 1960 and 1980. Because of this documentation, greater emphasis is now being placed on lime needs for this region in educational programs.

A unique approach was taken by workers in Virginia in the early 1970s (G.W. Hawkins, 1976, personal communication) in the use of soil test summaries to emphasize the need for lime. In the 5 years previous to 1973, lime use in Virginia had been declining, while use of acid-forming N fertilizer was increasing; steps were deemed necessary to prevent a serious low pH problem. The Extension soil fertility specialists developed a lime emphasis program employing soil test summary data that determined the increase in net income from proper use of lime and encouraged application of good liming practices. To determine net income for lime, these workers:

1. Worked with major crops at the county level.
2. Determined the crop land area.
3. Estimated the average soil type.
4. Determined the land area in each soil pH category.
5. Determined the lime needed to raise soil pH to 6.5 for each pH category.
6. Estimated the yield increase for each pH category.
7. Calculated the net income (increased profit minus cost of lime).

The estimation of average soil type was based on available data on average crop yields for different soils in Virginia. The percent of samples was translated into percent of hectares for each category. It was believed that this assumption could be made with some reliability when working with one crop in one county. An example of the use of soil test summary data for this program is given in Table 12-9. This illustration was made on a

Table 12-9. Calculating net income, 1973 Virginia Tech lime emphasis program for alfalfa (*Medicago sativa* L.), assuming potential yield for average soil of 9 Mg/ha.†

Soil pH	Percent of samples	Hectares	Total lime needs, Mg	Yield increase, Mg
0.0–4.9	1.4	450	6 040	5 080
5.0–5.4	5.6	1 810	18 245	14 220
5.5–5.9	13.4	4 340	31 585	9 720
6.0–6.4	27.0	8 750	39 190	9 800
6.5+	52.5	17 010	+0	+0
			95 060	38 820

Value of yield increase: 38 820 Mg @ $44/Mg = $1 712 000
Net income: $1 712 000 − $760 480 (cost of lime @ $8/Mg) = $951 520

† G.W. Hawkins, 1976, unpublished data.

statewide basis, although the actual summary material prepared for the program was done on a county basis by the Extension agronomist and local county agent. Along with emphasizing the need for lime by use of soil test summary information, additional county meetings were held where lime needs were discussed, and leaflets were extensively distributed on the merits of lime use. As a result of this program, there was a 56% increase in lime use in Virginia in 4 years.

Evaluating Soil Fertility Programs for Turfgrass

The Virginia Tech turfgrass program has used different types of summaries to convey information regarding soil fertility levels for encouraging the proper use of fertilizer and lime. Approximately 30% of the soil samples tested by the Virginia Tech Soil Testing and Plant Analysis Laboratory are submitted by homeowners and approximately 50% of these samples are for home lawns. Soil test summary data have been presented at turf educational meetings in four different ways. The following tables were prepared for a statewide program.

First, the general fertility status of the soil was discussed (Table 12–10). Examples of items emphasized were (i) the excessive amounts of golf greens that were testing "very high" in P where one might experience zinc (Zn) deficiency problems, and (ii) in the lawn establishment category, the number of

Table 12-10. Soil test P levels for five turfgrass categories in Virginia, 1983 (Donohue, 1984).

| Turf category | P soil test level | | | |
	Low	Medium	High	Very high
		%		
Putting greens	4	24	42	31
Tees	8	30	39	23
Fairways	4	32	51	13
Lawn establishment	21	36	27	16
Lawn maintenance	8	29	36	27

Table 12-11. Soil amendment needs for five turfgrass categories in Virginia, 1983 (Donohue, 1984).

Turf category	Percent needing fertilizer, lime†		
	pH	P	K
Putting greens	2	4	38
Tees	15	8	32
Fairways	11	4	11
Lawn establishment	48	21	35
Lawn maintenance	34	8	24

† Indicated by "low" P, K; pH <5.5.

Table 12-12. Soil amendment adjustment requirements for five turfgrass categories in Virginia, 1983 (Donohue, 1984).

Turf category	Percent requiring alteration of program†		
	pH	P	K
Putting greens	28	35	38
Tees	36	31	35
Fairways	21	17	16
Lawn establishment	59	37	38
Lawn maintenance	45	35	29

† Indicated by "low" or "very high" P, K; pH <5.5 or >7.0.

samples (21%) that fell in the "low" category where one would expect a deficiency if remedial action was not taken.

In the second table in this sequence (Table 12-11), an evaluation was made of the percent of samples requiring fertilizer and/or lime. This was indicated by samples that fell in the "low" category for P and K, and those with a soil pH <5.5.

In the third table in the sequence (Table 12-12), an indication of the percent of samples requiring alteration of the fertilizer and lime program was given. Changes in the program was deemed necessary if the P or K level was either in the "low" category where fertilizer was definitely needed or in the "very high" category, where excessive amounts could lead to nutrient tie-up or to imbalances. Alteration of the lime program was deemed necessary when the pH either fell below 5.5, when lime would be needed, or was >7.0, where trace element deficiencies might occur.

In the fourth table (Table 12-13), information was presented on the change in soil fertility status over time. One point highlighted in this table was the 2.5-fold increase in lawns needing K between 1971 and 1983. This was presumably brought about by the change in use from a "farm-type" fertilizer such as a 10-4.4-8.3 (10-10-10) to a primarily N-only fertilizer during this 12-year period. These are some of the uses that can be made of soil test summaries in an educational program.

In the previously mentioned survey on the value and use of soil test summaries, 22 of the 40 laboratories responding indicated that they used soil test summaries to evaluate fertilizer and lime recommendations, whereas 37 of the 40 laboratories used summaries to encourage proper use of fertilizer

Table 12-13. Change in soil fertility status for four turfgrass categories in Virginia from 1971 to 1983.

| | Percent needing lime, fertilizer† | | | | | |
| Turf category | pH | | P | | K | |
	1971	1983	1971	1983	1971	1983
			%			
Putting greens	13	2	3	4	32	38
Tees	53	15	5	8	6	32
Fairways	16	11	4	4	8	11
Lawns	36	35	9	9	10	25

† Indicated by pH <5.5; "low" P, K.

and limestone. It was indicated that soil test summaries were primarily prepared for soil pH, P, and K with fewer states summarizing for the secondary- and micronutrients. Soil test summaries, if used judiciously, can provide meaningful information on soil fertility levels and can be used to promote the proper use of fertilizer and lime. The increased use of computers in soil testing laboratories today, facilitates the preparation of soil test summaries and encourages their use in the educational program.

SUMMARY

Many soil testing laboratories prepare summaries of soil test results for their state or region. The value of these summaries in identifying fertilizer and lime needs depends, to a large extent, on the manner in which the data are summarized. Summaries of soil test results that are averaged over crops and different physiographic regions can give misleading information on the fertility status of the soil. When soil test summaries are separated according to crop by county within a physiographic region, data tend to be more meaningful, although it has been suggested that biases may be created by an indeterminate number of samples taken for troubleshooting purposes, the time of year the sample was taken, year-to-year variation in soil test levels, climatic differences within a region, land on short-term lease, and other factors. Each factor is reviewed individually in this chapter. Soil test summaries, which are rather extensively produced, are used only to a moderate extent throughout the USA to evaluate and promote the proper use of fertilizer and lime. Single-year data as well as multi-year data, which can describe trends in soil test levels over time, can be used to evaluate soil fertility status. Proper dissemination of this information can result in beneficial shifts in fertilization and liming for more economical crop production.

REFERENCES

Donohue, S.J. 1984. Virginia soil test summary for fiscal year 1983. Virginia Polytechnic Institute and State Univ. Ext. Pub. 452-004.

Donohue, S.J. 1985. Virginia soil test summary for fiscal year 1984. Virginia Polytechnic Institute and State Univ. Ext. Pub. 452-015.

Harrison, J.S., J. Byars, D.E. Peaslee, and K.L. Wells. 1983. Evaluation of farmer-submitted soil samples as an estimate of the fertility status of a county. Commun. Soil Sci. Plant Anal. 14:1181-1191.

Liebhardt, W.C., and M.R. Teel. 1977. Fluctuations in soil test values for potassium as influenced by time of sampling. Commun. Soil Sci. Plant Anal. 8:591-597.

Mahler, R.L., A.L. Halvorson, and F.E. Koehler. 1985. Long-term acidification of farmland in northern Idaho and eastern Washington. Commun. Soil Sci. Plant Anal. 16:83-95.

McCollum, R.E., and W.L. Nelson. 1954. How accurate is a summary of soil test information? Soil Sci. Soc. Am. Proc. 18:287-292.

Parker, F.W., W.L. Nelson, E. Winters, and I.E. Miles. 1951. The broad interpretation and application of soil test information. Agron. J. 43:105-112.

Peters, J.B. 1978. Lime requirement survey of north central Wisconsin soils and the effect of soil pH on corn yield and nutrient uptake. M.S. thesis. University of Wisconsin-Madison.

Thomas, G.W., and D.E. Peaslee. 1973. Testing soils for phosphorus. p. 115-132. *In* L.M. Walsh and J.D. Beaton (ed.) Soil testing and plant analysis. Soil Science Society of America, Madison, WI.